Being Human

Technology and animals often serve as the boundaries by which we define the human. In this issue contributors explore these categories as necessary supplements or as porous membranes which disturb the scaffolding of how the human is constructed. A lingering question throughout is whether we have ever been human or if such a category is a non-localizable ideal or perhaps a misnomer. In this collection of essays, internationally known theorists muddle the categorical boundaries such that animals and technologies become necessary components rather than limits for what it means to be human. They examine a range of subjects, including apophatic animality, critical media objects-to-think-with, biosemiotic insect resonances, the monstrous and horrific which dislodges our cultural animals, and the problem of thinking of animality as stupidity. Novels, films, digital objects, scientific laboratories, philosophical texts, animals on the road and in the fields serve as sites for inquiry. The result of these investigations is the spectral possibility that we are not the humans we make ourselves out to be.

This book was originally published as a special issue of *Angelaki*.

Ron Broglio is Director of Graduate Studies at Arizona State University, USA, and Senior Scholar at the university's Global Institute of Sustainability. He is author of *Surface Encounters: Thinking with Animals and Art* and *Technologies of the Picturesque*.

Frederick Young is faculty at the University of California, Merced, USA. He collaborates with conceptual artists and is working on a theoretical treatise on revolution, animality and technics. He is the author of *Toward an Ethics of the Political* and has published on Critical Theory and Art Practice.

Being Human

Between animals and technology

Edited by
Ron Broglio and Frederick Young

LONDON AND NEW YORK

First published 2015 by Routledge

2 Park Square, Milton Park, Abingdon, Oxon OX14 4RN
711 Third Avenue, New York, NY 10017, USA

Routledge is an imprint of the Taylor & Francis Group, an informa business

First issued in paperback 2017

British Library Cataloguing in Publication Data
A catalogue record for this book is available from the British Library

ISBN 13: 978-1-138-80786-0 (hbk)
ISBN 13: 978-1-138-05906-1 (pbk)

Typeset in Bodoni MT
by RefineCatch Limited, Bungay, Suffolk

Publisher's Note
The publisher accepts responsibility for any inconsistencies that may have
arisen during the conversion of this book from journal articles to book chapters,
namely the possible inclusion of journal terminology.

Disclaimer
Every effort has been made to contact copyright holders for their permission to
reprint material in this book. The publishers would be grateful to hear from any
copyright holder who is not here acknowledged and will undertake to rectify
any errors or omissions in future editions of this book.

Contents

CONTENTS

Citation Information

The chapters in this book were originally published in *Angelaki*, volume 18, issue 1 (March 2013). When citing this material, please use the original page numbering for each article, as follows:

CITATION INFORMATION

Chapter 7
Allison Hunter
Angelaki, volume 18, issue 1 (March 2013) pp. 99–106

Chapter 8
Insects and Canaries: Medianatures and aesthetics of the invisible
Jussi Parikka
Angelaki, volume 18, issue 1 (March 2013) pp. 107–120

Chapter 9
Tolstoy's Bestiary: Animality and animosity in The Kreutzer Sonata
Dominic Pettman
Angelaki, volume 18, issue 1 (March 2013) pp. 121–138

Chapter 10
A Global Cinematic Zone of Animal and Technology
Seung-Hoon Jeong
Angelaki, volume 18, issue 1 (March 2013) pp. 139–158

Chapter 11
Doing and Saying Stupid Things in the Twentieth Century: Bêtise *and animality in Deleuze and Derrida*
Bernard Stiegler
Angelaki, volume 18, issue 1 (March 2013) pp. 159–174

Chapter 12
Five Heraldic Animals (for Eduardo Kac)
Steve Baker
Angelaki, volume 18, issue 1 (March 2013) pp. 175–180

Chapter 13
After Animality, Before the Law: Interview with Cary Wolfe
Ron Broglio
Angelaki, volume 18, issue 1 (March 2013) pp. 181–190

Please direct any queries you may have about the citations to
clsuk.permissions@cengage.com

Notes on Contributors

Steve Baker is an independent writer, researcher and artist, affiliated to the University of Central Lancashire, Preston, UK, as Emeritus Professor of Art History. His most recent book is *Artist | Animal*. He is also the author of *The Postmodern Animal; Picturing the Beast*; and, with the Animal Studies Group, *Killing Animals*. His recent roadkill imagery has been reproduced and discussed in the journals Antennae, Art & Research, and Tierstudien, and shown in group exhibitions in London, New Orleans and Melbourne.

Ron Broglio is an associate professor in the Department of English at Arizona State University, Phoenix, USA, and Senior Scholar at the University's Global Institute of Sustainability. His research focuses on how philosophy and aesthetics can help us rethink the relationship between humans and the environment. He is author of *Surface Encounters: Thinking with Animals and Art* (2011) and *Technologies of the Picturesque* (2008).

Allison Hunter is Artist-in-Residence at Rice University, Houston, Texas, USA.

Seung-Hoon Jeong is Assistant Professor of Cinema Studies at New York University Abu Dhabi, UAE. He received Korea's Cine21 Film Criticism Award in 2003 and the Society for Cinema & Media Studies Dissertation Award in 2012. He has published on diverse filmmakers such as Werner Herzog, Peter Greenaway, and Apichatpong Weerasethakul, on major theorists such as André Bazin, Gilles Deleuze, and Jacques Rancière, and on critical issues such as the animal, multiculturalism, and catastrophe in cinema. His latest book is *Cinematic Interfaces: Film Theory after New Media* (Routledge, 2013).

Stephen Loo is an architect, artist and philosopher, and Professor of Architecture at the University of Tasmania, Australia. He has published widely on language, affect and the biophilosophy of the contemporary subject, posthumanist ethics and experimental digital thinking. His forthcoming edited book with Hélène Frichot is *Deleuze and Architecture*.

John Mullarkey is Professor of Film and Television Studies at Kingston University, London, UK. He has also taught philosophy and film theory at the University of Sunderland, UK, and the University of Dundee, UK. He has published *Bergson and Philosophy* (1999), *Post-Continental Philosophy: An Outline* (2006), *Philosophy and the Moving Image: Refractions of Reality* (2010), and edited, with Anthony Paul Smith, *Laruelle and Non-philosophy* (2012). He is currently working on a book entitled *Reverse Mutations: Laruelle and Nonhuman Philosophy*.

Marcel O'Gorman is an associate professor in the Department of English at the University of Waterloo, Canada, and Director of the Critical Media Lab. His published research, including *E-Crit: Digital Media, Critical Theory and the Humanities* (2006) and *Necromedia* (2015), is concerned primarily with the fate of the humanities in a digital culture. His most recent work investigates the "collusion of death and technology," a concept that he calls "necro-media." He is also a practising artist, working primarily with physical computing inventions and architectural installations.

Jussi Parikka is Reader in Media & Design at Winchester School of Art,University of Southampton, UK, and Adjunct Professor in Digital Culture Theory, University of Turku, Finland. He is the author of *Digital Contagions* (2007), *Insect Media* (2010), *What is Media Archaeology?* and *Anthrobscene* (2014). Recently he edited the online book *Medianatures*.

Dominic Pettman is Chair of Culture and Media, Eugene Lang College, as well as Associate Professor of Liberal Studies, at the New School for Social Research, New York City, USA. He is the author of *After the Orgy: Toward a Politics of Exhaustion* (2002), *Avoiding the Subject: Media, Culture and the Object* (2004), *Love and Other Technologies: Retrofitting Eros for the Information Age* (2006), *Human Error: Species- Being and Media Machines* (2011) and *Look at the Bunny: Totem, Taboo, Technology* (2013).

Daniel Ross is the author of the book *Violent Democracy* (2004) and director of the film *The Ister* (Black Box Sound and Image, 2004). He has written numerous articles on Bernard Stiegler and translated many of his works, including the books *Acting Out* (2009), *For a New Critique of Political Economy* (2010), *The Decadence of Industrial Democracies* (2011) and *Uncontrollable Societies of Disaffected Individuals* (2012).

Undine Sellbach is a philosopher, writer and performer, based in the School of Philosophy, University of Tasmania, Australia. Her publications explore the imagination in relation to notions of life, gender, instinct, ethics and the unconscious. With Stephen Loo, she is currently working on a monograph on psychoanalysis, ethics and the entomological imagination of childhood. Their performance-based work includes *Whirlwind of Insects in the Body of a Girl* for the Sexuate Subjects international conference at University College London (2010), *The Grasshopper Cabaret* for the Architecture-Writing symposium at KTH Stockholm (2012) and *Mistress O & the Bees* for the Kelly's Garden Curated Projects and the Sydney Biennale (2012).

Bernard Stiegler is a French philosopher, Director of the Institut de recherche et d'innovation, and founder of the School of Philosophy at Epineuil-le-Fleuriel, France. Since 1994 he has published some thirty books, including *Technics and Time* (1998), *The Decadence of Industrial Democracies* (2011), and *Uncontrollable Societies of Disaffected Individuals* (2012). His most recent major works are *Ce qui fait que la vie vaut la peine d'être vécue. De la pharmacologie* (2010); E*tats de choc. Bêtise et savoir au XXIe siècle* (2012) and *Pharmacologie du Front national* (2013).

Eugene Thacker is the author of a number of books, including *In the Dust of this Planet – Horror of Philosophy, vol. 1* (2011), and *After Life* (2010). Thacker teaches at The New School, New York City, USA.

Tom Tyler is Senior Lecturer in Philosophy and Culture at Oxford Brookes University, UK. His research concerns the use of animals, and the persistent expression of anthropocentric assumptions, within philosophy, critical theory, and popular culture. He is the editor of *Animal Beings* (2006), the co-editor of *Animal Encounters* (2009), and the author of *CIFERAE:ABestiary in Five Fingers* (2012).

Cary Wolfe's books and edited collections include *Animal Rites: American Culture, the Discourse of Species, and Posthumanist Theory* (2003), the edited collections *Zoontologies:*

The Question of the Animal (2003) and, most recently, *What is Posthumanism?* (2010) and *Before the Law: Humans and other Animals in a Biopolitical Frame* (2012). He has also participated in two multi-authored collections: *Philosophy and Animal Life* (2008), and *The Death of the Animal: A Dialogue* (2009. He is founding editor of the Posthumanities series at the University of Minnesota Press and founding director of 3CT: Center for Critical and Cultural Theory, at Rice University, Houston, USA.

Frederick Young is faculty in the Merritt Writing Program at the University of California, Merced, USA. Many of his publications are concerned with animality and post-structuralist theory as well as moving towards art practice. He has collaborated with several conceptual artists. Currently he is working on a lengthy theoretical treatise on the question of revolution, animality and technics. With Linus Lancaster, he is collaborating on the same topic using art practice as an interdiction for a new "politics of art."

Agamben's anthropological machine serves as a useful way to begin thinking of the relationship between humans, animals, and technology. Because the anthropological machine decides each and every time in favor of the human, it is the figure for that which includes and excludes, that which divides inside from outside and so forms a community of those within the anthropological circle of the human and those who remain on the outside looking in:

> Insofar as the production of man through the opposition man/animal, human/inhuman, is at stake here, the machine necessarily functions by means of an exclusion (which is also always already a capturing) and an inclusion (which is also always already an exclusion). Indeed, precisely because the human is already presupposed every time, the machine actually produces a kind of state of exception, a zone of indeterminacy in which the outside is nothing but the exclusion of an inside and the inside is in turn only the inclusion of an outside. (37)

The human functions as a "transcendental signifier produced through the various and multifarious instances of its own failure" (Oliver 8). The empty signifier functions as the difference which makes a difference between claimants as (in)human. As Agamben points out, at the margins humans may be judged inhuman, as with the Jews during the Holocaust, and the inhuman may be judged as human as in the proto-human ape-man or "missing link" of early natural history.

Agamben points to Heidegger as the last philosopher to believe "that the anthropological machine, which each time decides upon and recomposes the conflict between man and

EDITORIAL INTRODUCTION

ron broglio

WHEN ANIMALS AND TECHNOLOGY ARE BEYOND HUMAN GRASPING

animal, between the open and the not-open, could still produce history and destiny for a people" (75). Dasein has a "historicity" and a placeness (a *polis*) and the human as Dasein has these because of our access to the open, our ability to think the question of being and distinguish between being and beings. Animals captivated by their environments – their Umwelt – are unable to get outside their worlds and have a look around; they cannot see beings as such and distinguish this "as such" as a question of being. The result for Heidegger is the famous claim that animals are "poor in world" while humans are "world forming" (*Fundamental Concepts* §42, 176). Animals are trapped within their world by their captivation while humans build a world rather than live merely trapped within one. This lays the groundwork for Heidegger's later work such as "Building, Dwelling, Thinking" where technology, thinking, and human comportment or dwelling are intertwined. We'll return to this alignment of concepts in a moment, but for now let us return to the anthropological machine which makes possible the division between those who build and dwell within their world and those who merely inhabit their Umwelt.

Agamben's concern in the anthropological machine is humans and humanness. He is interested in which humans count as human and which – like the Jews in Nazi Germany – are excluded from the circle of humanity. Such exclusion licenses treating the excluded as less than human, as animal. (This particular problem is well developed in Seshadri's recent book *HumAnimal: Race, Law, Language*.) As for the animals themselves and even the animality of man, these remain disarticulated from Agamben's concerns for the social community. Agamben is not concerned with the outside as outside; he is not concerned about the animals as such but only as they function as a "means of an exclusion (which is also always already a capturing)." The animals are included inside the system as a figure of the outside. Or using his terms, they are captured by the system as that which is excluded from the social life of humans. As Dominic LaCapra explains in his critique of Agamben:

> There is a sense in which, in Agamben's own discourse, animals in their diversity are not figured as complex, differentiated living beings but instead function as an abstract philosophical topos similar in certain respects (perhaps even functioning as a displacement of) the *Muselmann* [...] Both "the" animal and "the" *Muselmann* function as avatars of the radically "other" (albeit, expectably, an other that is also within the self). (LaCapra 166)

And this other, while radically other, is still within the system of the anthropological machine which inscribes, describes and decides the difference between the inside and outside. The animal is the symmetrically other to the human within Agamben's figuration of the machine system of insides and outsides, open and non-open (Wolfe 24).

Could there be an outside to this symmetry, an outside beyond the outside that is figured by and for the anthropological machine? In what is perhaps the most promising moment within *The Open*, Agamben points to Heidegger's admission that animals may well have their own open but it would be something to which humans are foreclosed from understanding:

> This question now leads us toward the distinction we tried to express by talking of man's *world-forming* and the animal's *poverty in world*, a poverty which, roughly put, is nonetheless a kind of wealth. The difficulty of the problem lies in the fact that in our questioning we always and inevitably interpret the poverty in world and the peculiar encirclement proper to the animal in such a way that we end up talking as if that which the animal relates to and the manner in which it does so were some being, and as if the relation involved were an ontological relation that is manifest to the animal. The fact that this is not the case forces us to claim that the *essence of life can become accessible only if we consider it in a deconstructive* [abbauenden] *fashion*. But this does not mean that life represents something inferior or some kind of lower level in comparison with human Dasein. On the contrary, life is a domain which possesses a wealth of openness with which the human world may have nothing to compare. (*Fundamental Concepts* 255)

It is exactly this "wealth of openness" which provides a possible outside beyond the dialectic of human-animal as figured by the anthropological machine. It is an outside beyond Agamben's concern for a machine that decides on humanness. Rather, such a wealth of openness is foreclosed to humans. Such a wealth is available to "life" and what LaCapra calls "animals in their diversity." This space of the animal open provides a way for thinking of the assemblage animal–human–machine without deciding in advance in favor of the human.

By invoking the wealth of openness, Agamben (following Heidegger) turns the open into a problem not only for animals but also for man. Man cuts himself off from his animal nature *but* he does so only to tend to the open and let beings be as such. Yet, as Agamben declares deftly, animals "know neither beings nor nonbeings, neither open nor closed, it is outside of being; it is outside in an exteriority more external than any open, and inside in an intimacy more internal than any closedness" (91). So, to let the animals be would mean to keep them outside of our knowing. Such outside would have to include the animal man and so we risk ourselves, suspend ourselves. What then would man and animals be who are outside of history and outside of the anthropological machine which separates the human from his animal nature?

There is a difficulty in thinking this wealth of openness to which we do not have access. Agamben calls this inaccessible realm a "zone of non-knowledge" or an "aknowledge." Because it is by definition foreclosed to culture, Agamben brackets out this space and time. At best, it becomes a figure of the messianic banquet and end of history which he uses as a bracketing figure in the opening and closing chapters of *The Open*. The end of history haunts the book but never intervenes or disturbs his focus: the constitution of the human (who is the agent and subject of history).

What is missing, what remains outside our knowing in this wealth of openness, is an animal phenomenology. To take this outside of the outside seriously would mean a hospitality in thought in the zone of non-knowledge or aknowledge. Such a hospitality would challenge what constitutes thinking. Being outside and an aknowledge, we cannot know this world of the animal and its wealth of openness. Put simply, we cannot know what it is like to be an animal from the animal's point of view – as Thomas Nagel so aptly displays in "What is it like to be a bat?"; however, we do get glimmers of this outside as it pushes against culture. If culture is that which defines itself over and against animality – as Agamben's

dialect displays – then an animality of the outside would disturb the axis of the humans and animals established by the anthropocentric machine.

It would take a revolutionary intervention to disrupt this axis of human and animal and to entertain the sphere of animal openness. As I've posited elsewhere, an animal revolution is such an untimely overturning. Its time is not that of history which is a mark of human time. Instead, animals in their openness live in a temporality that is virtual to ours. This is to say, the time of animals is an-other time, one parallel to our own. Occasionally these temporalities collide; history meets the time of the animal. In such moments, the wealth of openness and the time of the animal erupt like an event on to our history.

Such events are multitude but by way of example we can consider the rather recent event, the case of Santino the chimp. Prior to examining Santino's story and the animal techne and worlding he proposes, consider the case of the fictional ape, Red Peter from Kafka's short story "A Report to an Academy." Red Peter is taken from his life in West Africa and placed on a ship for Europe. While aboard he learns to be human by watching the sailors. It is not that he wants to "ascend" to humanity but simply, while being barred up, he wants a way out: "Up until then I had had so many ways out, and now I no longer had one." His way out is to become like the humans – to act like everyone else. Red Peter begins his report by explaining that he in fact cannot report his life as an ape: "your experience as apes, gentlemen – to the extent that you have something of that sort behind you – cannot be more distant from you than mine is from me." To become human is to no longer be ape. Red Peter cannot translate between these worlds so different in kind. Kari Weil explains the ape's predicament:

> Language is at the core of Kafka's critique of assimilation as a process that gives voice only by destroying the self that would speak. What is the self, Kafka's story asks, if it has no memory of its past and no means of

representing it? Must that (animal) self be a blank slate for others to write upon? Or might there be some other source of selfhood in his body, some physical locus where memory may be stored and known? (6)

The Academy with all its cultural apparatus is foreclosed from the aknowledge of the ape's wealth of openness. Indeed, Red Peter (before being Red Peter) had "so many ways out" and now has only one, acting civil. Civility is, he admits, not freedom, but it provides him movement and livelihood. What the Academy cannot access is the life of the ape.

Enter then the case of Santino. Santino the chimpanzee likes to throw rocks at visitors to the Furuvik Zoo in Sweden just north of Uppsala. Early in the morning he combs the grounds of his compound to find handy stones and piles them into a cache along the bank next to where the visitors will gather. If need be, he chips off some concrete and breaks it into a reasonable throwing projectile. It is not uncommon for chimps and apes to throw things at zoo guests but what is strikingly uncommon is his forethought, a characteristic we have reserved for humans. "Planning for a future, rather than a current, mental state is a cognitive process generally viewed as uniquely human. Here, however, I shall report on a decade of observation of spontaneous planning by a male chimpanzee in a zoo," and so begins Mathias Osvath's report to the Academy in *Current Biology* (120). It is one of the first well-documented cases of non-instinctual and non-habituated forethought in a nonhuman.

Forethought is a characteristic fundamentally aligned with being human, so much so that Heidegger uses the concept as the scaffolding for building Dasein. Visualizing a future allows humans to plan and build, to make objects function as equipment toward future ends, and to fashion technology. In response to Santino's forethought, Osvath wrote his report and the zoo keepers castrated Santino as a mode of control over his behavior and future plans. Red Peter's way out is to become human and assimilate in "a process that gives voice only by destroying the self that would speak." By

contrast, Santino bears witness to another world and provides another sort of voice – albeit with rock in hand. Osvath compares Santino's craftsmanship to that of hominins:

> The behaviors also hint at a parallel to human evolution, where similar forms of stone manipulation constitute the most ancient signs of culture. Finds as old as 2.6 million years suggest that hominins carried and accumulated stone artefacts on certain sites, presumably in case of future need planning. (121)

Of course, this is neither human evolution nor "ancient signs of culture" but rather a present-day event in which an outside animality throws a nonhuman culture against our own cultural expectations.

A standard definition of techne – one put forth by Heidegger in "Question Concerning Technology" – contrasts with *physis*. While *physis* is "the arising of something out of itself," techne is the arising or revealing of something as brought forth by another such as an artisan or craftsman (317). Santino gives us an outside that cannot be assimilated into the human–animal axis of the anthropological machine because the wealth of openness is beyond the grasp of such an anthropocentric system. This exteriorization by means of chimp technology points to an unassimilated uniqueness in animals and their life worlds, their phenomenologies. In *Before the Law*, Cary Wolfe develops what the realm of nonhuman cultures means in terms of philosophy:

> Though there is no doubt a vast qualitative difference between the developing modes of human exteriorization and "grammatization" and those of other species – a point on which both Derrida and Stiegler would agree – the animal behaviors and forms of communication we have been discussing are "already-there," forming an exteriority, an "elsewhere," that enables some animals more than others to "differentiate" and "individuate" their existence in a manner *only possible* on the basis of a complex interplay of the "who" and the "what," the individual's "embodied enaction" (to use Maturana and Varela's phrase) and

exteriority of the material and semiotic technicities that interact with and rewire it, leading to highly variable ontogenies, complex forms of social interaction, individual personalities, and so on. (76)

Other individuals with other grammatization and other tools hurled at us from "elsewhere" remind us that our world is not enough nor our knowledge sufficient to make sense of non-human openness.

Santino's case is but one of many incidents where humans, animals, and technologies collide in a way that disturbs human worlding and points us beyond ourselves. Consider the following as a small sample. (In a related strain, see "Vengeful Tiger, Glowing Rabbit" by Randy Malamud.)

Radioactive Board: In the case of the Chernobyl nuclear disaster, the devastation wrought by technology upon the bodies of humans has been forgotten by many who spin the narrative of technological progress as a mode of bettering society. The better we build, the better our lives will be – such is the thinking that begets a building and spawns a way of dwelling. Yet, invisibly lurking in the ground and in plants is the castoff refuse of a forgotten disaster and a dwelling otherwise. Ingested by boars, radioactivity returns but now with tusks and a wild disposition. Radioactive boar roam the streets of German towns in Bavaria. The return of the repressed: technology gone awry is now nature without domestication. Godzilla as a monstrous return of a repressed nuclear disaster in Japan may be a fiction, but these boar are more than awkwardly made film creatures. The boar have found the weakness of our strength. If our strength and happiness is our ability to forget and repress, then they have used this forgetting as a blind spot by which to invade our cities. (Hawley)

Commando Sheep: Sheep in the Pennines, UK have learned how to roll over the cattle grid barrier to get to greener pastures. Witnesses say that the "sheep have perfected their version of the commando roll" and are now destroying local gardens. (Wainwright)

Technological Octopi: The curious and seemingly gregarious female octopus of a California aquarium tugged on a valve that allowed hundreds of gallons of water to overflow its tank and flood the exhibition floors. Aquarium spokeswoman Randi Parent says no sea life was harmed by the flood, but the brand new, ecologically designed floors might be damaged by the water. Octopi incidents are ramped. One has been spotted carrying a turtle shell as a shield, another creates a kabuki-like set of camouflage routines, others are studies by astrobiologists for possible likeness to alien bodies and alien intelligence. (Miraglia)

Cougar from elsewhere: Fish and Wildlife Services of the U.S. announce the extinction of cougars in New England. July 26th 2011 a cougar shows up in Connecticut having walked 1,500 miles from the Black Hills of South Dakota. Apparently this cougar had been reading the Fish and Wildlife Services reports and wanted to repopulate the area or just prove the government surveyors wrong. *The New York Times*'s David Baron writes: "if a cougar can walk from South Dakota to Connecticut, a cougar could show up anywhere" (Baron). No place is safe or immune from the possibilities of animality and the animal revolution. The virtual of the animal lurks and can erupt and tear the fabric of the social surface at any moment.

Each incident is a call from elsewhere which disrupts Dasein's designated unique relationship to the call of/toward being. These calls of the wild from animals that "know neither beings nor nonbeings" (as quoted from Agamben above) undo a singular call (*anruf*) that awakens Dasein to itself and its destiny (Heidegger, "Letter on Humanism" 245). The following collection of essays by leading authors in critical theory and animal studies assimilates incidents, calls, and disruptions and in most cases such an event is occasioned through a mode of *production* by which thought thinks differently and at its limits.

Disruption is evident in what John Mullarkey cites as the fundamental horror in cinema. In "Animal Spirits: Philosomorphism and the Background Revolts of Cinema" he explains:

What will the animal revolution look like? Horrifying cinema outlines it for us: its many stories, told through a play of sound and light, offer us glimpses of various future programmes. It tests theories by developing new logics of visibility and invisibility, of what it means to belong or be expelled, to think and decide, to be vulnerable and to suffer, to force oneself "into a community," to be deemed alive.

Cinema has the power to produce couplings between technological apparatus and animal Umwelt to provide a glimpse of the outside beyond human conceptualization. What Mullarkey proposes is a non-philosophy (following François Laruelle) which undoes philosophers' appropriations of animals for their own ends or what he calls "philosomorphism" which shapes the animal to the ends of human thought.

Production as a mode of thinking animality, techne and the human is most evident in Marcel O'Gorman's essay "Speculative Realism in Chains: A Love Story." He provides a story of assemblages of desire between objects in the world, techne, equipmentality and the human. For O'Gorman, we are in a world and worlding that implicates our relation to objects such that desire becomes part of our thinking of things. Given our intertwining, our chained relation, to objects, he proposes

> a new economy of care in our technocultural system, [by which] philosophers would do well to intervene directly in that system's mode of production, namely digital production, offering alternative models for technological invention that draw attention sideways, to the complex entanglements of human and nonhuman things.

While Mullarkey looks at what cinema produces and O'Gorman spurs thought by experiments with the chains of techne and equipmentality (most notably his Mikado bike), Stephen Loo and Undine Sellbach utilize the laboratory of Jakob von Uexküll. They explore how this forefather of biosemiotics establishes a unique mode of production by which "genuine biological investigation entails a certain willingness, on the part of the scientist, to evoke to the 'mind's eye' what is forever inaccessible to our physical senses – the radically different spatial, perceptual, temporal and affective worlds of other animals." Loo and Sellbach then note that Uexküll's famous work "A Stroll through the Worlds of Animals and Men" is subtitled "A Picture Book of Invisible Worlds." They develop how Uexküll and then they themselves leverage the genre of the picturebook to "creating an aesthetic medium in which new and unforeseen sensibilities might emerge."

Tom Tyler's "New Tricks" follows on the heels of his recent theoretical work *Ciferea: A Bestiary in Five Fingers*. Tyler retains the influence of critical theorists but turns to how video game production reveals new ways of thinking animal worlds. Much as Loo and Sellbach leverage the storybook, Tyler looks at how "the Smellovision technology" in the game *Dog's Life* "mounts a genuine challenge to customary thinking about the canine." Through handling a canine avatar, players extend their way of thinking into nonhuman realms and inhuman sensory abilities.

Continuing his work on life, vitalism, and horror, Eugene Thacker's contribution, "Apophatic Animality: Lautréamont, Bachelard, and the Bliss of Metamorphosis," looks at Lautréamont's *Maldoror* as a work that

> itself seemed like an animal, a teratological anomaly composed of bits and pieces, a *corpus* left unfinished or untended. In contrast to the textuality of the animal so frequently found in literary representation, *Maldoror* seems to put forth the animality of the text – composed of multiple tendrils, leaping off the page, devouring the reader.

Through reading this work that inspired surrealists and situationists, we discover an inhuman phenomenology, "animality on a phenomenological plane, as a 'vigorous poetry of aggression.'" Thacker pushes such phenomenology to or beyond its limit as *"animality is no way reducible to animals"* but rather "the text moves from reproduction to production, from representation to presentation" that infects those whom it touches.

Production and event reveal something new in Jussi Parikka's essay "Insects and Canaries: Medianatures and Aesthetics of the Invisible." Here Parikka extends work he has done in *Insect Media* by looking at how insects communicate to us the state of affairs in ecologies that remain otherwise invisible to humans. The inhuman insects make palpable a world beyond our grasp: "the animal is not only an object of concern but is itself a surface of registration, storage media and a signal of the processes concerning pollution and waste." Global honey bee colony disorder haunts and informs this essay that extends from species extinctions (via Ursula Heise) to art that highlights the in/visible ecological worlds – what Loo and Sellbach call in their essay on Uexküll "the radically different spatial, perceptual, temporal and affective worlds." Parikka explores the video art installation by Lenore Malen as "A slowly progressing multiplication of viewpoints" which produces "the becoming-animal of perception."

Dominic Pettman moves from his analysis of posthumanism, animality, and techne in *Human Error: Species-Being and Media Machines* to another mode of production: thinking animality through the bestiary. In "Tolstoy's Bestiary: Animality and Animosity in *The Kreutzer Sonata*," Pettman gives voice to the "figures deployed in Tolstoy's story – sometimes explicitly, sometimes through allusion – to describe the ultimately *in*human dispensation of love." The accumulation of bestiary incidents amid the social discourse of *The Kreutzer Sonata* produces a porous membrane by which human autonomy and civility unravel.

In "A Global Cinematic Zone of Animal and Technology," Seung-hoon Jeong moves through a number of contemporary films that entwine humans and animals. The anthropocentrism in *Project Nim* and Herzog's *Grizzly Man* reveal the difficulty of thinking animality. It is with *Tropical Malady* that Seung-hoon Jeong finds a mode of thinking such entwining. The film provides "an inert interface to a world larger than human, the Virtual immanent in the Actual in Deleuze's terms."

I am very pleased to include in this issue on animality and techne Bernard Stiegler's essay "Doing and Saying Stupid Things in the Twentieth Century: *Bétise* and Animality in Deleuze and Derrida." Stiegler addresses the problem of how we become human or, as O'Gorman explains in his essay which deploys Stiegler's recent work, how we learn to care and through care become individuals. In this essay, Stiegler develops Simondon's notion of individuation to read Deleuze on stupidity [*bétise*]:

> This is the question of *individuation* and *dis-individuation*. If we are *able* to be stupid, it is because individuals only individuate themselves from out of *preindividual funds* from which they can *never break free: from out of which, alone, they can individuate themselves,* but within which they *can also get stuck, bogged down,* that is, dis-individuate themselves.

While Stiegler takes on the problem of individuation, Cary Wolfe in his recent work *Before the Law* asks how humans and other animals become defined by the linguistic and social mechanisms of language, reason, and law. Wolfe ruminates on the biopolitical frame where "before" is both antecedent to and under the diktat of the law. Critically, he asks where biopower might break with the enframing of the biopolitical:

> bodies are enfolded via biopower in struggle and resistance, and because those forces of resistance are thereby produced in specifically articulated forms, through particular *dispositifs*, there is a chance – and this marks in no small part Foucault's debt to Nietzsche (as both Esposito and Deleuze point out) – for life to burst through power's systematic operation in ways that are more and more difficult to anticipate. (Wolfe 32)

Wolfe works through not only the theoretical apparatus of animality but also the comportment of animals – from factory farming to pets to play where animals do not simply "react" but "respond" and give a glimpse into animal worlding. My interview with Wolfe addresses not only his recent book but the

University of Minnesota Press series "Posthumanism," which he edits, and his insight into directions in which the field of posthumanism is moving.

I'm pleased to include in this issue artwork by Allison Hunter and Steve Baker. These works display the complexity of multiple worlds – human and animal – entwined and bound with our technological worlding. Hunter's work from her *Honeycomb* series speaks to colony hive collapse and complements Jussi Parikka's essay on insect media as well as the insect worlding of Uexküll's storybook laboratory. The layers of surfaces in the bee works and the dragonfly work remind the viewer of the surface encounters between worlds and the organic and inorganic assemblages that are wrought. Steve Baker's "Five Heraldic Animals (for Eduardo Kac)" works in dialogue with a number of essays in this collection. On the road to human progress, we meet (meat) the friction of animality. Our own technological capacities (envisioned in the motorized wheel) attempt to elude our animality – to transport us elsewhere – and yet fatalities do occur, as Baker has captured. More abstractly, the surface and fur of the animal bodies shown here create patterns that echo the patterns depicted in Hunter's honeycomb images. Looking at the honeycomb and fur we seek patterns of meaning but always within these there lies embedded the immanence of animality.

I hope the reader will take these essays' experiments, productions, and queries as ways of thinking a "wealth of openness with which the human world may have nothing to compare." Uexküll believed such experiments would lead to "worlds strange to us but known to other creatures, manifold and varied as the animals themselves" (5). The essays here expand what it means to be human in multiple ecologies.

bibliography

Agamben, Giorgio. *The Open: Man and Animal.* Stanford: Stanford UP, 2003. Print.

Baron, David. "The Cougar behind Your Trash Can." *The New York Times* 28 July 2011: A27. Print.

Hawley, Charles. "A Quarter Century after Chernobyl Radioactive Boar on the Rise in Germany." *Spiegel Online International* 30 July 2010: n. pag. Web. 1 Aug. 2012.

Heidegger, Martin. *The Fundamental Concepts of Metaphysics.* Bloomington: Indiana UP, 1995. Print.

Heidegger, Martin. "Letter on Humanism." *Basic Writings.* Trans. David Farrell Krell. New York: HarperCollins, 1993. 213–66. Print.

Heidegger, Martin. "The Question Concerning Technology." *Basic Writings.* New York: HarperCollins, 1993. 311–41. Print.

LaCapra, Dominic. *History and its Limits: Human, Animal, Violence.* Ithaca, NY: Cornell UP, 2009. Print.

Malamud, Randy. "Vengeful Tiger, Glowing Rabbit." *Chronicle of Higher Education* 23 July 2012: n. pag. Web. 24 July 2012.

Miraglia, Niki. "Octopus Opens Valve and Floods Aquarium." Associated Press 26 Feb. 2009. Web. 22 Jan. 2012.

Oliver, Kelly. "Stopping the Anthropological Machine: Agamben with Heidegger and Merleau-Ponty." *PhaenEx* 2.2 (2007): 1–23. Print.

Osvath, Mathias. "Spontaneous Planning for Future Stone Throwing by a Male Chimpanzee." *Current Biology* 19.5 (2009): 120–21. Print.

Seshadri, Kalpana Rahita. *HumAnimal: Race, Law, Language.* Minneapolis: U of Minnesota P, 2012. Print.

Uexküll, Jakob von. "A Stroll through the World of Animals and Men: A Picture Book of Invisible Worlds." *Instinctive Behavior: The Development of a Modern Concept.* Ed. Claire H. Schiller. Trans. D.J. Kuenen. New York: International UP, 1957. 5–80. Print.

Wainwright, Martin. "Pennine Spot where Sheep Won't be Fenced In." *The Guardian* 30 July 2004: n. pag. Web. 1 Aug. 2012.

Weil, Kari. *Thinking with Animals: Why Animal Studies Now?* New York: Columbia UP, 2012. Print.

Wolfe, Cary. *Before the Law: Humans and Other Animals in a Biopolitical Frame*. Chicago: U of Chicago P, 2012. Print.

Ontology, or first philosophy, is not an innocuous academic discipline, but in every sense the fundamental operation in which anthropogenesis, the becoming human of the living being, is realized [...] From the beginning, metaphysics is taken up in this strategy: it concerns precisely that meta that completes and preserves the overcoming of animal physis in the direction of human history.

Agamben, The Open *79*

Muybridge seemed to be racing against the imminent disappearance of animals from the new urban environment. Distinct from the stillness of photography, cinema added the possibility of electric animation [...] The advent of cinema is thus haunted by the animal figure, driven, as it were, by the wildlife after death of the animal.

Lippit, Electric Animal *185–86, 197*

john mullarkey

ANIMAL SPIRITS
philosomorphism and the background revolts of cinema

introduction: the cinematic animals that we always were

This essay follows two lines, one that we can call (for now) "cinematic," the other "philosophical," towards an intersection located in "the animal." It is not that either domain is drawn to the animal for its own intrinsic worth, but rather that they are pushed there by their own "passion for the Real" (to use Badiou's phrase), or at least in order that they might create certain "reality effects" through the animal (Badiou, *Le Siècle* 75ff.). Be it the bleak picture of "bare life" drawn by Agamben, or the more positive image of the "animal that therefore I am" depicted by Derrida, philosophers of various hue have shown increasing interest in the idea of the animal as both a normative category (Derrida, Agamben) and a metaphysical one (as when Badiou depicts Deleuze's philosophy as one of "the Animal" in contrast to his own of "Number" – Badiou, "Review" 55). One question that arises from this is whether or not these myriad philosophies mediate the animal for their own philosophical purpose by seeing it as only an instance of aporetic *différance* (Derrida), proliferated becoming (Deleuze), bare life (Agamben), or even the very model of "bad philosophy" (Badiou). Too often, I would argue, the animal does indeed appear as an avatar of one or other such philosopheme.

And yet any reduction of the non-philosophical presence of the animal to that of being a proxy for *différance*, rhizomatics, bare life, or whatever else gains its apparent force in part only by ignoring other aspects of the animal that are placed in the background, namely those that do not fit (or resist) the philosopher and his/her favoured philosophemes. The background revolts. It is not simply that Derrida's "*animots*" are not like Heidegger's "*weltarm*" animals, or Agamben's, or Deleuze's, but that the ontological leverage acquired through reference to the animal (as when Deleuze sees "the agony of a rat [as] [...] the constitutive relationship of philosophy with nonphilosophy") always has an ineliminable remainder, some parts of which are occupied by other rival philosophical positions on the animal, other parts of which escape Theory altogether (Deleuze and Guattari, *What is Philosophy?* 109).

However, while no one philosophy may acknowledge its own reductive abuse of the animal as such (at least if it is serious in its own intent), recourse to the model by which *cinematic* animality is both articulated and "backgrounded" (especially in horror cinema) reveals, as we will see, an isomorphism between two thwarted desires. The very turn to the lives, deaths, and transformations of animals (into humans – and of humans into animals) in order to enhance the philosophical and cinematic Real also transcends that same move on account of an inherent irreducibility, one that exposes it as only the latest gesture by which thought tries to commandeer reality.

There remains, however, the very live question as to whether every use of animals by Theory (perhaps philosophy especially) must amount to an anthropomorphic abuse of some sort, be it positive or negative, inflationary or deflationary (as John S. Kennedy for one has argued).[1] Is every philosophy of the animal a type of what might best be called "philosomorphism" – refracting the animal through an image of itself? Perhaps. Yet, in virtue of the resistance of the Real of animality to such uses, there will also be a possible contrary movement of zoomorphism, understood here as a possible *non-human philosophy*, whereby the various philosophies that try to capture the animal in their epistemological, ontological, or metaphilosophical categories are nonetheless resisted in such a way that it is the very idea of philosophy itself that must also be reshaped in order for it to say anything significant about the animal. Adopting the stance of the "nonphilosopher" François Laruelle, we ask whether, in its (unacknowledged) attempts to shape the animal in its own image, philosophy also succeeds in refracting itself. There is a circularity of philosophical explanations (as Laruelle finds in all philosophy in fact) that is linked to this sub-species of anthropomorphism: be it positive or negative, inflationary or deflationary, such philosomorphism is indeed resisted by the Real of animals by mutating or morphing what counts as philosophy – thought, reason, logic – too. And cinema – which is a kind of animal thought in all of us – provides just such a mutation.

It is also worth asking why the animal should be such a potent resource *now*, especially for philosophy. The contemporary "animal turn" in Theory appears timely, I would argue, in that the animal marks two kinds of limit for both cinematic and philosophical thinking that are respectively internal and external. The potency of the (film) image of the animal, I'll argue, indicates the point of its indiscernibility with film itself as animal thinking. The representational and affective power of cinema is clearly immense, and many theories – psychological and philosophical – have been offered to explain its effects on us: Freudian, Cavellian, Cognitivist, and so on. The answer outlined here, however, is that the force of cinema simply *is* the power of the animal that we (always) are when we think in images, or when images think immanently within us. The capacity that film has to make us see and think differently is less about its representations and more about its materiality as an animal mode of thought. Linda Williams, following Carol Clover, has famously written of the three "body genres" of cinema that most obviously disturb our flesh (melodrama, horror, pornography) and, certainly, when we let the image arouse our tears, screams and genitalia, we do seem to approximate a Pavlovian dog becoming over-excited by stimuli.[2] Yet there are also more complex responses, which are neither thoughtless flesh nor disembodied reflections, but affective thoughts, seeing-thoughts, that are all the more potent because they are imagistic. These images are not any less the animal-thinking-in-us, however, nor are they either base or inhuman: they are simply where a powerful, animal form of non-human thinking resides.

Of course, various different models of film thinking (and even film philosophising) have been proposed by a number of film theorists and film-philosophers alike, be they illustrative (film translating verbal thoughts into images) or native (film thinking – and communicating – through its indigenous audio-visual medium). I have examined many of these models elsewhere, especially those purporting that film thinks in its *own* way (for instance, Stanley Cavell and Gilles Deleuze).[3] Such a claim leads one to question what a cinematic logic would look like: could it ever approximate the precision of *bivalent* logic, for example? The challenges facing such a thinking in images, indeed the challenge for both montage (the relations between images) *and* cinematography (the relations within images) could be pictured thus: what is the visual difference between "This and that" vs. "This or that" vs. "This not that" vs. "If this, then that"? In other words, how can one visualise the difference between conjunction, disjunction, negation, or conditionals? One image simply following another does not seem to be able to capture these precise differences. And yet there are alternative logics (and with that different forms of philosophical "rigour" and "consistency") available when we look closely at the specifics of the production, exhibition, and reception in camera distance, frame and composition, the dissolve, inter-cutting, the tracking shot, zoom, multi-screen, focus, lighting, background and foreground, projection speed, screen luminescence, etc. In this essay, it will be the thinking of horror film (or the horrific *mode* found in *all* film) that will show us a new reality effect that involves the *visualisation and sounding of the background*.[4] It concerns those putatively inert entities that film allows

to emerge from the background as living, a background(ed) life itself that, by refusing to be ignored any longer, may also reform (or morph) what we mean by philosophy itself. The emergence of the cinematic background horrifies and "revolts" (in both political and affective senses), and this is its new kind of thinking.[5]

philosomorphs of the human philosophers

As we will see in the second part of this essay, the lives and deaths of animals have always played a role in creating the reality effects of cinema. However, amongst the following four philosophers too – Derrida, Deleuze, Badiou, and Agamben – the same life and death has equally had real effects on their definitions of metaphysics and politics, and even of thought and philosophy. Jacques Derrida clearly believes that his work on the animal opens up new meanings for philosophy, as seen, for instance, when he speculates that when "the animal looks at us, and we are naked before it. Thinking perhaps begins there" (Derrida, *The Animal* 29). Even his critique of "biological continuism" (Singer, Midgley et al.) as the basis for an animal advocacy can be seen as predicated upon his own theoretical inclination towards difference over identity. Derrida is not interested in erasing any demarcating lines (to create a continuum) but prefers rather to multiply them on all "sides," thickening and thinning the lines as he goes. Hence, there is no "Animal" (which may or may not include the human-animal), but only *animals* plural, or, in the French, "*animaux.*" But *animaux* is a French homophone with Derrida's neologism "*animot.*" This phonic identity subtended by written difference allows him to play upon the written French plural for animals and the French for word (*mot*), thereby revealing once more (as he did with *différance*), how writing determines linguistic meaning through difference (Derrida, *The Animal* 41, 47–48, 55). Donna Haraway's criticism of Derrida in all this, though, is that he still remains too negatively anthropomorphic – caring more for *his* response to the gaze of his "little cat" and its *hypothetical* response, as though the cat had no *real* response of its own. How the actual cat actually regards him, without any reference to human interest (projected or not), is left unnoticed amidst all of Derrida's "worrying and longing" over the cat's response (Haraway 20). He lacks the curiosity (of a cat) to enquire after its own inner life and needs, preferring to remain within a relationship built from one-sided pity and mutual exclusion. Likewise, Jonathan Burt highlights Derrida's attraction to the negative, and ultimately to death, as a mark of his work that also mediates the animal.[6] Death refracts his thinking about animals, and the animal – as mute, as other – becomes an effect of (his) morbid writing:

> This anachronism in the contemporary philosophization of the concept "animal" has significant consequences for the gaps it opens up, almost in spite of itself, between metaphysics (theology, ontology) and language on the one hand, and the animal's specific place(s) in the contemporary world on the other. The morbidity of the former runs deep [...] The ease with which such thinking lends itself to a grander metaphysics that seems far removed from the realities of human–animal relations need not, in principle, be objected to. These traditions have long constituted much of our thinking about other beings, worlds, nature, and so on. But this conceptual version of the "animal," with all the connotations of muteness and melancholy that underpin its symbolization, is a consequence of this thinking rather than constitutive of it, despite the claims of the theory that sacrifice is a foundational act in the cultural categorization of beings. The animal is, in other words, a writing effect that latches onto a more generalized, and inflated, concept of otherness. (Burt, "Morbidity" 158–59)

It is a philosomorphic attribution that is at work here – using the animal as one more philosopheme, an avatar of vulnerable, silent difference. Indeed, our "crimes" against animals, according to Derrida, prolong the violence that begins in speech (*logos*), extends itself through patriarchy (phallocracy), and completes itself by implicating our entire species in its war, aptly now re-dubbed "*carno-phallogocentrism*" (Derrida, "'Eating Well'" 112). So, when Derrida turns to Bentham he does not invoke suffering as an *ability* (*à la* Singer) – or indeed any other property/power such as language, consciousness, or reason – but as a fundamental *passivity*. Suffering is a vulnerability, a finitude, that we share (in a non-continuous way) with the animal, and that makes us animals.

Finitude, vulnerability, death, difference, thinking. The animal is here more a philosomorphic creature than any other kind. For, conversely, Haraway's call to attend to the animal's own response inevitably falls foul of the usual Cartesian demand – how does she know that she's any more right than Derrida for *not making that attribution*? Yet it is the very inevitability of this pitfall that led Mary Midgley, for one, to disavow the whole anthropomorphism debate and remove it from the charge sheet against philosophy and science: we all do it, all the time, with other life-forms and with each other too, yet we don't coin names for our "illegitimate" paedomorphism, gerontomorphism, andromorphism, gynaecomorphism, or even plutomorphism and paupermorphism.[7] The question is not whether we do it (as it is ubiquitous and unavoidable), but *in what manner* do we do it?

So, can the animal respond at all, *in its own way*, to the philosopher's call for a response (from it) and (our) responsibility (to it)? For Derrida, there might be a way, but:

> it would not be a matter of "giving speech back" to animals but perhaps of acceding to a thinking, however fabulous and chimerical it might be, that thinks the absence of the name and of the word otherwise, as something other than a privation. (Derrida, *The Animal* 48)

Absence as more, not lack, not as privation, but as differentiation. This is not a differentiation *of* any thing (*ousia*, presence, *logos*) but for itself as an end in itself (ethically *and* metaphysically). Derrida's philosophy of difference is perfectly suited to adopting the animal as one more face for its own philosopheme. A "chimera," needless to say, is an animal composed from parts of multiple other animals – a creature of montage, or assembly. Pack animals, animals as assemblages of other, smaller (larval) animals, is precisely what marks the Deleuzian philosomorphic approach:

> We think and write for animals themselves. We become animal so that the animal also becomes something else. The agony of a rat or the slaughter of a calf remains present in thought not through pity but as the zone of exchange between man and animal in which something of one passes into the other. This is the constitutive relationship of philosophy and nonphilosophy. (Deleuze and Guattari, *What is Philosophy?* 109)

Pity, of course, would be a reactive relation that does neither the dying animal nor thinking philosopher any good. Derrida's pity for the animal is a morbid response that Deleuze's vitalist thinking cannot stomach. The ethics of this relationship is strictly affective, concerning forces, or degrees of motion and rest. It is also asymmetric – not in the Levinasian mode of being *for* the other but in the naturally aristocratic mode of being for the greater good of a healthier, more joyful relation: it is *the relation* that is the end in itself:

> Man does not become wolf, or vampire, as if he changed molar species; the vampire and werewolf are becomings of man, in other words, proximities between molecules in composition, relations of movement and rest, speed and slowness between emitted particles. (Deleuze and Guattari, *Anti-Oedipus* 107)

Of course, Luce Irigaray famously took Deleuze and Guattari to task for insisting upon a "becoming-woman" that would incorporate both "molar" women *and* men within a greater molecular process, to the neglect of any real identitarian politics (whose achievements had only recently been hard-won by actual women).[8] Might not the same be said of this "becoming-animal" in as much as, *pace* their assertion that Deleuze and Guattari think and write "*for* animals" and even become

animal "*so that* the animal also becomes something else," it is really only *human* becoming that counts for them? The animals' part in this pact is only a means to an end: "the becoming-animal of the human being is real, even if the animal the human becomes is not" (Deleuze and Guattari, *A Thousand Plateaus* 238).[9]

On the one hand, it is doubtless that this commendation of process over product is part of an anti-representationalist, anti-mimetic tendency in his thought: in order to "be the Pink Panther," one must imitate and reproduce "nothing"; but like that big cat, one "paints the world its colour, pink on pink; this is its becoming-world" (*A Thousand Plateaus* 11). Deleuze's "animal philosophy" (as Badiou has dubbed it) is a philosophy of life that, like its forbear in Bergson, is utterly immanentist – there is nothing outside of becoming to become. Molarity, be it of the human or the non-human, is not to be imitated; even more, it *cannot* be imitated (hence, what looks like reproduction is always a new production, a productive repetition). Leaving aside the paradox of having to *commend* what everything and everyone is already anyway (in a philosophy of absolute immanence),[10] the fact remains that there are kinds of animal that Deleuze prefers over others in this all-encompassing molecular becoming: domesticated (pitied) and individuated (molarised) animals are unhealthy – this being the motive behind the infamous proclamation that "anyone who likes cats or dogs is a fool" (*A Thousand Plateaus* 239). Likewise, state animals (the lions, lambs, horses, and unicorns of empires, myths, and religions) are to be disavowed. It is the demonic or pack animal that is the Deleuzian favourite – rats, wolves, cockroaches. So, qua animal *becomings* (rather than the becoming-animal of humans), the true animal is always a multiplicity (as in a wolf pack) and a process (every such pack is a *wolfing*).

On the other hand, however, qua becoming-animal, what should be (at least) a two-sided process involving healthy, active becomings for all participants invariably profits only the human. For example, in the section on the "War Machine" in *A Thousand Plateaus*, the nomad invention of the stirrup involves a human becoming-animal within a "man–horse–stirrup" assemblage. Similarly, the bridle used in masochistic practice is itself derivative of a human "becoming-horse" (*A Thousand Plateaus* 399, 260). It seems that, as a state animal, the horse can only serve a Deleuzian becoming when participating in human enhancement. Obviously, Deleuze is fully alerted to the danger here of "becoming-animal" manifesting as a mere imitation of a "molar" image (*playing* "horsey"), or an aspect of horseness in the abstract or as a part-object festishism: this is supposed to be a process of re-creating certain speeds and slow-nesses – metamorphosis over representation (or metaphor). Yet Deleuze seems oblivious to the politics of his examples, horses in particular. How can a stirrup involve any kind of *becoming* for the horse (remembering that all Deleuzian becomings must be *active*, not *reactive*)? How can the stirrup be part of any thing affirmative? A domesticated horse may take pleasure in ful-filling some of its functions, but that is only because it has been broken in, made into a sad animal. The technology itself is reactive – a poli-cing and controlling of life rather than its further production. And, as for the agony of a dying rat or slaughtered calf, the technologies involved in "pest" control and factory-farming can hardly be deemed life-affirming *from the point of view of the animal lives involved*.[11] There seems little place for the animal's own genuine "becomings" in the alter (animal)-side of a human becoming-animal. What of its or their "becoming-human"? Is that always and only a token of anthropomorphism – a matter for projection, sentimentality, and pity? Where is the example of human and animal change as a real co-becoming, such as the wasp and orchid enjoy in *A Thousand Plateaus*, that is, an active, joyful "deterritorialization" for *both* (*A Thousand Plateaus* 293)?[12] As things stand, the animal, demonic or domestic, is a phi-losophical myth in Deleuze's thought, a philoso-morph that must play a lesser part in the architecture of becoming, for any such process is always evaluated from the perspective of the human alone. Indeed, such is the immanence

of Deleuzian thought, it is even arguable that the animal *must* remain outside genuine becom-ing if, as he says, its agony and slaughter are constitutive of a zone of exchange between human and animal as well as philosophy and non-philosophy. The dying animal is the outside of thought, the shock to thought, and so the non-philosophical as such for Deleuze. To be allowed its own life and vital becoming would be to allow it too much philosophy, too much thought for itself, too much *anthropos*. And yet this very withdrawal of the relationship, or its suddenly unilateral bias, is itself a token of the philosomorphic use of the animal (Nietzschean, Spinozist, Bergsonist) pervasive throughout Deleuze's thought.

Indeed, Alain Badiou's criticism of Deleu-zianism as a philosophy of the animal (over his own of Number) spies just such a possibility in Deleuze and actualises it (contra Deleuze's own intent) as an animalisation of the human that thereby strips humanity of value. Yet Badiou's intervention is not only to run Deleuze's animals in a new direction but also to protect his own avowedly humanist project from any possible animalist extension. The real threat is to Badiou's *universalist* and *extensionalist* politics of emancipation. Given that, according to *Being and Event*, no inten-sive quality can be legitimised that might restrict the scope of political change (an Event), it follows that the *greater* scope of pol-itical vitalism promises to undo any humanist agenda within such a universalism. How can some parts of the universe be kept out of poli-tics in order to keep the programme *human*? How can they be backgrounded so that the human alone figures in the foreground? What *cannot* be defined as "human" if one is not allowed to define the human at all intensionally, that is, through some or other actual or poten-tial quality – species-being, gender, race, nationality, civilising power, etc?

Badiou's concept of universalism (be it in art, science, politics and love) has a mathematical basis – *everything can be counted* (pure quan-tity, not quality, counts). However, just as Hei-degger, in *Being and Time*, decided to approach the question of Being through *Dasein* as the most appropriate method because only humans ask the question of the meaning of Being (despite initially arguing against confusing Being as such with *any* particular beings), so Badiou too must justify his anthropocentric ontology in the most parsimonious way possible: through mathematical ability. Though every thing can be counted, because only humans *can think infinitely, that is to say, can do mathematics* (in a sense so defined by Badiou as to exclude any animal facility for mathematics), only humans can enjoy political subjectivity. "Humans" (of a certain kind) count because they alone (supposedly) can count (despite all empirical continuities between human and animal abilities that demonstrate the contrary).[13] The finitude that Derrida sees as what animal and human share is here echoed in reverse for Badiou, for he emphasises the infinite power of human thought *over* its animal, embodied finitude. This is the one intensional property that Badiou must willingly opt for, as well as appropriately define, so as to ensure that his humanist agenda is saved from his universalist programme, and that the concept of the human does not ooze into the non-human. Here, though the animal is playing a negative role (rather than the ostensibly positive role it plays in Deleuzian thought), it remains crucial to Badiou's project that it *keeps* to that role, that it plays its philosophical part without resisting, that it conforms to the philosomorph designed for it.

Indeed, Badiou's philosomorphism is openly Platonic: when he does explicitly describe animals, he prefers their Forms over their specific instances. And, as with Deleuze, it is horses that loom large in this mediation. Badiou's *Logics of Worlds* opens with the Platonic Idea of the Horse as it runs from prehistorical cave painting through to Picasso. Contra Antisthenes' remark to Plato – "I do see some horses, but I see no Horseness" – Badiou contends that it is *nothing but Horseness* that we see:

> Nevertheless, I contend that it is indeed an invariant theme, an eternal truth, which is at work between the Master of the Chauvet cave and Picasso. Of course, this theme does not envelop its own variation, and everything we have just said about the meaning of the horses remains correct: they belong to essentially distinct worlds. In fact, because it is subtracted from variation, the theme of the horse renders it perceivable [...] We simultaneously think the multiplicity of worlds and the invariance of the truths that appear at distinct points in this multiplicity. (Badiou, *Logics of Worlds* 18)

With continued regard to this Platonism and politics of the Idea, in his earlier *Ethics* Badiou asks whether man is a "living animal or immortal singularity," and answers his own question by asserting that man is "an animal whose resistance, unlike that of a horse, lies not in his fragile body but in his stubborn determination to remain what he is" (Badiou, *Ethics* 11). Unlike a horse, humans rebel against domination. And they do it alone, winning power through violent revolutionary overthrow rather than democratic, gradualist reform (or pitying advocacy). Consequently, the founding of the RSPCA in 1824 by Richard Martin or the publication of Peter Singer's *Animal Liberation* in 1975 could not be seen as true political events for Badiou. He rejects the "contemporary *doxa*" of the "humanist protection of all animals" because, firstly, animals cannot count like adult humans can count (neither can human infants or some humans with specific disabilities, but let us not complicate things too much). But also, secondly, every political event in the name of emancipation and equality must come extensionally, from the "bottom up," such that advocacy ("protection" as well as representation) *cannot* be a true conduit for a political event (again, we will have to ignore the fact that the examples of revolution cited by Badiou – French 1789, October 1917 – involved some representatives acting on behalf of the proletariat).[14] So, while we might still pertinently ask whether equality does always come exclusively from the "bottom up," we might also ask what an animal revolution might look like, given that the *appearance* of a true revolution (qua political event, at least on Badiou's account) cannot be anticipated through any extant situation. How might those pushed into the background, the discounted, return into vision when such an event is unforeseeable in principle, or, at best, can only be glimpsed? How do animals revolt, or rather, what would a revolt and return (from *revolvere*) look like?

Though appearing more ambivalent in his speciesism than Badiou, Giorgio Agamben too (in his work on Heidegger in *The Open: Man and Animal*) is alarmed by "the total humanization of the animal [that] coincides with a total animalization of man" (Agamben 77). As Kelly Oliver notes, "in some passages in *The Open* Agamben seems to accept a Heideggerian abyss between man and animal, an abyss that Agamben suggests is not wide enough in Heidegger's thought" (Oliver 232). The threat to this protective abyss is the double-sided "anthropological machine" that results, in Heideggerian terms, in "a monstrous anthropomorphization of [...] the animal and a corresponding animalization of man" (Agamben 58). A monstrous anthropomorphism, born in the modern era with nineteenth-century biologism. But the machine has a longer history to it than just this, given its role of creating social inclusion through a posited "missing-link" that is neither properly human nor social, but animalised and "bare." Yet the machine works in various ways, not just one:

> On the one hand, we have the anthropological machine of the moderns [...] it functions by excluding as not (yet) human an already human being from itself, that is, by animalizing the human, by isolating the nonhuman within the human: *Homo alalus*, or the ape-man. And it is enough to move our field of research ahead a few decades, and instead of this innocuous paleontological find we will have the Jew, that is, the non-man produced within the man, or the *néomort* and the overcomatose person, that is, the animal separated within the human body itself. The machine of earlier times works in an exactly symmetrical way [...] here the inside is obtained through the inclusion of an outside, and the non-man is produced by the humanization of an animal:

> the man-ape, the *enfant sauvage* or *Homo ferus*, but also and above all the slave, the barbarian, and the foreigner, as figures of an animal in human form. Both machines are able to function only by establishing a zone of indifference at their centers, within which – like a "missing link" which is always lacking because it is already virtually present – the articulation between human and animal, man and non-man, speaking being and living being, must take place. Like every space of exception, this zone is, in truth, perfectly empty, and the truly human being who should occur there is only the place of a ceaselessly updated decision in which the caesurae and their rearticulation are always dislocated and displaced anew. What would thus be obtained, however, is neither an animal life nor a human life, but only a life that is separated and excluded from itself – only a *bare life*. (Agamben 37–38)

The machine creates anthropomorphs (animals become human), but also, dangerously, theriomorphs (animalised humans): the political dangers of abusing the latter are clear to Agamben (as they are to Badiou), but, as Oliver also points out, Agamben seems indifferent to the dangers for animals stemming from both versions of the machine, anthropomorphic and theriomorphic. But that is because, for Agamben, the animal *already resides within the space of the missing link: the fictions of both the animalised human and the humanised animal are both – in terms of philosophical worth* (reasoning power, language skills, political subjectivity, and even the ability to die properly, rather than simply cease to exist) – *possible types of real animal.* Because such purities (reason, language, subjecthood, death) are proper to actual, real, human life alone – "the truly human being." The impure product of any splicing of the truly human being with the animal results in and belongs with animality. So Agamben is disingenuous when he argues that bare life is "neither an animal life nor a human life," for it is much closer to the former in value. Akira Lippit puts the kind of speciesism at work here in a clear light. Agamben's animals are too dumb to die – and any facility they may have for it will have to be palmed off on the pre-linguistic powers of film (or something like it):

> If, according to the strained logic of Western metaphysics, the animal cannot die – to the extent that death is seen as an exclusive feature of subjectivity and is reserved for those creatures capable of reflecting on being as such in language – then the death of the animal in film, on film, marks a caesura in the flow of that philosophy of being. The animal dies, is seen to die, in a place beyond the reaches of language. (Lippit, "Death of the Animal" 18)

As Jonathan Burt glosses this idea, the animal never dies because it is constantly re-animated, repeating "each unique death until its singularity has been erased, its beginning and end fused into a spectral loop." In that manner, cinema keeps the animal "alive"; away from the dialectics of language, it cannot undergo a "proper death" (Burt, "Morbidity" 165). Animals, in terms of *philosophical value*, lack all perfection: any putative perfections in speed, grace of movement, strength, etc. are always, *pace* Nietzschean thought, an imperfection of sorts. And, like other humanists before him, Agamben must place the animal within this Christianised version of the philosomorph – in the horror of mixture. In other words, the bareness of bare life *belongs properly to the animal* (as a human stripped bare of its distinguishing qualities). Yet it is precisely within the *horror* of this mixing of the species where we might see where the political threat of an animal not being an animal might arise. This would be its revolt against philosomorphic thought.

notes on non-human philosophising

It is not that we ask to avoid all forms of philosomorphism altogether – as a sub-species of anthropomorphism, so long as there is more than oneself (that solipsism is false), then it will always be the case that some kind of leap of faith, act of charity, or hermeneutical stance is needed to understand one's other, whatever or whoever "it" may be. Anthropomorphism can be a very useful methodological tool in ethology.[15] What is too often the trouble with anthropomorphism, however, is that it is one-way and partial. What is required is a two-way and absolute anthropomorphism: this would not only involve projecting some of the molar aspects of human behaviour onto another being, alive or inert, in a unilateral fashion, but also the *absolute* leap of attributing *everything* to the other in a "principle of charity" without end, as well as introjecting from that other all that one can in the same movement. *Anthropos* is thus expanded, *morphed*, in and through the *non-*human (but this would be in Laruelle's positive sense of "non-" as broadened, mutated).[16] An absolute anthropocentrism reshapes both the *subject* (*anthropos*) as well as the *project* (animal). It is a bi-lateral, two-way pro-jection that is also an intro-jection.[17] The expansion is the charitable act, the leap of faith. It is also to "give" the benefit of the doubt, a gift and benefit that comes before all forms of representation (stances, leaps, gift-giving, and even Cartesian doubts).[18]

In which case, rather than reject other forms of philosomorphism (Derrida's, Deleuze's, Badiou's, or Agamben's) as misrepresentations (as if, finally, now, we could hope to represent things correctly), if we splice them together and review them as equal, immanent parts of a whole we also mutate what counts as philosophy just as much as philosophy mutates the animal to its own ends. This would be a chimerical philosophical animal that is only as real as the number of "eyes" it encompasses (as Nietzsche once described, with an appropriately monstrous metaphor, in his own non-relativist perspectivism).[19] Such an absolute or complete philosomorphism also mutates philosophy through the animal. And this would thereby entail rethinking what counts as thinking and vulnerability (Derrida), thinking and non-philosophy (Deleuze), thinking and political subjectivity (Badiou), thinking and humanity (Agamben).

Even so, why should the axiom for this practice involve rethinking these "thinking ands" (as if that was a presuppositionless starting point for philosophy) when we already have an alternative means of reviewing a set of human–animal chimera that is equally affective, yet less specifically focused on human forms reasoning? By this I mean, of course, the monstrous forms of cinematic montage, especially those seen in horror film – such will be our model for an expanded "thinking" (or reasoning, analysing, arguing, questioning, etc.). After all, in Noël Carroll's work, for example, we have the most interesting account of cinematic art-horror which argues that the chimerical and bareness of life are part and parcel of the monster at the heart of every horror film. The monster, to be truly horrific, must be "repulsive and abhorrent" (Carroll, "Film" 39). Such creatures are "impure," "categorically hybrid," compounding normally opposed elements: life and death – vampires, zombies; or human and demonic – devils, anti-Christs, and so on. But they also compound the animal and the human in werewolves, human flies, cat people, etc. Such horrifying hybridity defies "our conceptual schemes," horror monsters being "impossible" or "anomalous" entities (Carroll, "Why Horror?" 34, 35, 37, 39).

Admittedly, Carroll's approach will be rather too cognitivist for many looking for a proper alternative way of reviewing philosophy (being too obviously party to the conceptualist side of things), yet a more embodied approach still fits this account too (after all, the monster not only offends our logic but also our taste – it is repulsive and abhorrent). Just as Deleuze–Spinoza say that we do not yet know what a body can do, so in horror we are shown the many ways a human body can be mutated into non-human forms. And, for Deleuze, this morphing occurs not only in cinema but *as cinema*, whereby "the animal has lost the organic, as much as matter has gained life" (Deleuze, *Cinema 1* 51).[20] As Anna Powell writes in *Deleuze and Horror Film*, "philosophical thought does not have to be abstract or transcendental in nature. The horror film experience offers a particular quality of thought" (Powell 154). Just as the arrival of new cinema technologies (like 3D) are also often occasions that remind us that film viewing is a material sensation, so too can we be reminded by new

kinds of horror film that *all cinema begins as horror, that is, in the mode of the horrific no matter its nominal genre* – and this is quintessentially so as it shows us the human and non-human, the living and the inert, spliced together. The question here will be whether this also shows us *where philosophy begins* as well – in the horrific (at the animality of the human) as much as in wonder. If this is the case, seeing and feeling the horror of the human expanded into the non-human will be a new kind of thought and philosophy too – and so a complete philosomorphism.

animal-cinema/background-horror: what is the monster trying to show us?

A number of things need to be said when turning to the images of animals in film and the animal imagery of film. Burt notes well that "audiences often respond differently to animals or animal-related practices than they do to other forms of imagery." The animal image is a "form of rupture in the field of representation," such that the normal suspension of disbelief does not work for animals on screen as it does for humans. The animal here becomes a "tabula rasa" that, via developing cinematic technologies, offers us access to allegedly hidden realities through posited (technomorphic) non-human ways of seeing reality – animal point-of-view shots (Burt, *Animals* 11, 21, 53).

It is incontrovertible, of course, that animals have always been an essential part of our cinematic experience, being there at the very start of film in the proto-filmic experiments of Muybridge and Marey. The first film stars included the superstar dogs of the 1920s, such as Strongheart and Rin Tin Tin, and the first films often showed spaces where animals were on display: in arenas (horse races, bullfights, circuses), city streets (parades, transport, ceremonies) and zoos. It is noteworthy that cruelty towards animals was often involved in these spectacles: animals appearing in film often entailed some really horrific event for them. Even Muybridge staged the killing of a buffalo in order to have a "guarantor of authenticity" (Burt, *Animals* 22, 43). Animal suffering and vulnerability were equally there in the early "actualities," such as the Edison Company's *Electrocuting an Elephant* (1903), or the works of the Surrealists (*Un chien andalou*, 1929). Subsequent cinematic history displays a persistent recourse to pained and dying animals in films ranging from the sensationalist – *Cannibal Holocaust* (1980); to the artful – *Apocalypse Now* (1979), *Heaven's Gate* (1980), or *Caché* (2005): each record of violence serves to heighten the "authenticity" of their spectacle.[21]

Going further, there is also a long history in cinema of humanising the animal (anthropomorphism) and animalising the human (theriomorphism), through hybrids of animal and human beings (werewolves, insect-men, lizard men), or animal and human behaviour, as in feeding (vampires, zombies, cannibals) or politics (invading aliens, ostracising "freaks"). And horror film itself, in its origins and concepts (genres), can always be seen rethinking its internal relations (between filmic elements like sound, editing, performance, and so on) and external relations (in terms of such basic questions as "is horror ethical?," "what is horror?," or even "what is cinema?"). Horror shows this, it demonstrates this. It splices humans and animals together, through story, special effects, and its very technological basis.

Turning to the etymology of horror's "monsters," the question as to what the monster is trying to show us could simply be this – that it can "show" at all, that it has an image it can show. What horror (or the horrific) does is play with images in order to show continuities and discontinuities between human and nonhuman images such that cinema itself, as a constant sensory mutant, always borders on the horrific (at its best). Paul Wells moves in this direction when he says that "'shapeshifting' is a chief determining concept at the heart of the horror film, whether this be in the metamorphoses of the monster or the context in which it exists" (Wells, *The Horror Genre* 34); while Paul Coats goes even further in *The*

Gorgon's Gaze when writing that "the horror film becomes the essential form of cinema, monstrous content manifesting itself in the monstrous form of the gigantic screen."[22] We would add that not only scale but every weird thing that cinema does also reveals the potential horror, or wonder, when perceiving unnoticed parts of the world.

Linda Badley cites Clive Barker's story "Sons of Celluloid," which has a "monster that feeds on the projected emotion of motion picture audiences" (Badley 39). But it is more than the audience's *emotions* that are at stake here, it is every sensation; for as Steve Shaviro notes, "cinematic perception is primordial to the very extent that it is monstrously prosthetic. It is composed, one might say, of the unconscious epiphenomena of sensory experience" (Shaviro 30). Even in realist horror, such as David Lynch's theriomorphic *Elephant Man* (1980), what is truly monstrous and horrific reveals itself to be cinema itself. Lars Nowak writes:

> Merrick is [...] presented as being the offspring of an animal. By being labeled "the elephant man" in the freak show, he is transformed into a hybrid of man and elephant. This animalization is elaborated upon by the film itself, in which Merrick is identified with several fictitious animals [...] the intertwinement of Merrick's genesis and that of film draws an analogy between the two. It suggests that, just as Merrick is not his parents' legitimate offspring, but a monstrous deviation, film is also not photography's legitimate descendent [*sic*], but a monstrous one. (Nowak 72, 77)

So again: film is horrific in its origins – the *Arrival of a Train at La Ciotat* makes this plain. Film makes things live, it animates them (or rather, it re-animates those that have become dead to our eyes – it is the great *re-animator*). *Cinema makes an animal, an animate, of all*. Indeed, Alan Cholodenko has argued that all cinema *is* animation: "not only is animation a form of film, film, all film, film 'as such', is a form of animation [...] Put simply, for us animation is the first, last and enduring attraction of cinema, of film"

(Cholodenko 1). As Deac Rossell has also observed, before the cinema of narrative, before even the "cinema of attractions" was prevalent, the earliest film exhibitions astounded people less for their foregrounds of moving people (already seen long ago in magic lantern shows), than for the *background movement of things normally left unnoticed*, such as the shifting leaves in the trees.[23] Astounding, or horrific. It is not that the horror is in the story qua narrative genre, or in the presence of the monster, or in the performances, or any other element of the *mise-en-scène*: it is potentially in *all* of these contents in as much as film can make things weird and horrific in all sorts of small ways. It is in the *how*, how these contents are shown, how they appear (or make themselves appear) out of the background. The how of film, when horrific, is *a way of showing and sounding* that renews the optical and the auditory.

When the horrific is seen as this modal process, then, and not as a genre, we get beyond any idea of generic essence, or rather, we forward the horrific as the essence of non-essence, as the *process of hybridisation*: creating new ideas of the generic, the general, and genre. Horror then becomes a kind of (philosophical) conceptualisation and generalisation, whereby concepts mix and mutate and everything oozes, becomes "weird."[24] The best horrors fail to conform to a "standard cinematic form" therefore: things are slightly too close, or too far away, askew, too loud, or too quiet; they renew the optical and auditory with weirdness, such that when a human body, for example, is seen up close too much, it is no longer human at all but becomes a mass of pitted matter. For an anthropocentric phenomenology such as Merleau-Ponty's, these normative terms like "too close" or "too far" indicate that the "true" meaning of the thing itself, in this case the human living body, only becomes manifest within a certain range of distances (Merleau-Ponty 302).[25] But to a non-human philosophy – such as is found in horror cinema – it reveals new lives that normally remain in the background.

do look back: animist spirits and the revolt of the background

In Japanese horror films ("J-Horror") like *Ringu* (1998) and *Ju-On* (2003), then, it is *the image itself that is monstrous*: black and white, silent, and emerging from the screen's background – like the Lumière brothers' train. And what is this image trying to show us if not that it is alive? Film – best known through the old names, *Biograph*, *Vitascope*, *Bioscop* – animates whatever it shows: it is the re-animating process as such. *Ringu* is a film that plays with images in order to show ties connecting the living and the inert – in photographs, videos, telephones, all optical and audio media. And it is Japanese animism that lies at the heart of not just J-Horror but the very horror, and wonder, of cinema itself.

This animist element must not be underestimated.[26] As Koji Suzuki, the author of the *Ringu* books, said in a 2005 interview:

in America and Europe most horror movies tell the story of the extermination of evil spirits. Japanese horror movies end with a suggestion that the spirit still remains at large. That's because the Japanese don't regard spirits only as enemies, but as beings that co-exist with this world of ours. (Balmain ix)

Of course, there is also the inorganic life of things that "animates the inanimate" in German expressionist horror such as *Nosferatu* (that Deleuze himself points to).[27] But in *Ringu*, *Ju-On* and much contemporary Asian horror (such as *Shutter* or *Pulse*), it is television sets, photocopiers, and photographs that come alive, such that the animism is of the optical machine itself.[28] Stephen Asma puts this is Freudian terms: *Ju-On* and *Ringu* are explorations of "the uncanniness of mundane objects, motions, and lights" (Asma 318 n. 24). Yet we should not let the horrific descend into the Oedipal pit: the uncanniness here does not concern a return of a repressed familial past but a direct, optical non-human present, emerging from our background into our foreground.

The return of the oppressed. These animate and animal spirits "co-exist" with us, though normally only as consigned to the background. This can be seen, for example, in the US remake of *Ju-On*, which plays fast and loose with the original multiple family-centred storylines, yet is extremely faithful to its key, horrific, optical situations. Jay McRoy gets this right when looking at the work of Takashi Shimizu (the director of *Ju-On*):

> by vacillating between limiting what we see and revealing the objects of our fear in groundbreaking ways that separate him aesthetically from other directors, Shimizu creates a text that may well alter forever the way that some viewers process cinematic horror. By frequently relegating frightening images to the extreme edges of the frame, thereby investing them with the power of a fleeting, yet troubling figure glimpsed peripherally but never completely, Shimizu artfully manipulates the audience's gaze, creating the impression that we may have just witnessed a flash of something disquieting – as if from the corner of our collective eye. (McRoy 181)[29]

J-Horror goes beyond the intermingling of molar humans and animals that characterises Western horror such as *The Wolfman*, and instead mixes the living and the inert, the optical foreground and background, centre and periphery.[30] But it does so in order to show, to demonstrate as only film can, that what we *saw* as dead is actually alive and always was: animism is not a hauntology. J-Horror is not object-oriented horror (as a kind of "spooky physics," as Gerald Edelman might label it); it is background-oriented, looking back and "thinking backwards" or in "reverse" (undoing the Kantian anthropocentrism and philosomorphism of the Real). Or rather, it is when the background will no longer be ignored and reasserts itself as a kind of optical and audio subject, a kind of human that was never, in-itself, an object at all, but only seemingly so through our disregard. This is its revolt, its return that forces us to look back at it.[31]

conclusion: towards a non-philosophy of animal revolt

With the return of the (evental) subject in Continental philosophy through Badiou's work, one might have thought that there could be an emancipatory politics for animal subjects that is not top down and anthropomorphic but bottom up and "animalist," or even vitalist. Indeed, one might even ask whether such a politics could be one of Badiou's non-philosophical truth conditions, or explore its connections with the "nonphilosophy" Deleuze relates to the suffering animal, or Agamben's definition of "first philosophy" in terms of the "becoming human of the living being." Can such a suffering or (for Derrida) "vulnerable" animal be a political subject, can it too "become human"? Badiou is adamant that it cannot: quoting one of his masters, Sartre, he is dogmatically against the possibility of animals being political subjects; after all, "every anti-communist is a dog."[32] Hence, his *Logics of Worlds* concludes by returning to its introductory attack on what he portrays as our current age of vitalism or "democratic materialism," which Badiou decries as merely living a life "without Idea" (as opposed to a "living for an Idea" that he claims to find in materialist dialectics).

However, other critics, such as Katerina Kolozova, make exactly the opposite point by emphasising a "solidarity" with the suffering, animal body as a non-philosophical basis for politics.[33] Real, horrific, pain is the basis for a true "political universal," a "lived revolution." And, in extrapolating from the non-philosophy of Laruelle, with her we too can posit a non-human politics and thought that can show us what a bottom-up politics of the animal might look like – *through cinematic thought*.[34] The "non-human" is not a negation but an extension, a mutation of the human into the animal and vice versa. What Agamben describes as the "non-man produced within the man" is not the opposite of first philosophy (or "anthropogenesis") but its fulfilment in the non-philosophical extension of the human, a "human-without-humanism."[35] In

Forme et objet, Tristan Garcia cites the view of anthropologist Eduardo Viveiros de Castro that what counts as human is entirely perspectival, such that "in the world *for us*, an animal species amongst others, we are 'human', but in the world of the jaguar, another animal species, the jaguar is 'human' for itself" (Garcia 240). This is also Laruelle's absolute democracy, which democratises what democracy means, and who it stands for, in the non-human and non-person (or "in-person"), always expanding the sense of these terms by mutation. Whether any of these non-philosophical and non-humanist strategies escape the charge of being anthropomorphic or even another form of philosomorphism remains moot. Indeed, the real question may be whether they actually need to escape it at all. An absolute philosomorphism refracts and mutates both the (non-human) animal and (human) philosophy into something new.

In "Thought Creatures," Eugene Thacker writes:

> Political vitalism is the sovereign correlative to the kind of vitalism-without-content that Agamben critically speaks of. It is the "animation" of sovereign power as legitimized by a higher order, or by a vital principle that is not immanent to the field within which politics possibly takes place [...] Thus the question: What would we have to "abandon" in the political-as-human in order to think beyond this political vitalism and consider a different kind of "vital politics"? (Thacker, "Thought Creatures" 328)

Our answer to this question is that, far from abandoning the "political-as-human," we expand it in a *genuine* universalism. Absolute anthropomorphism is projection for its own sake, a charitableness without end, or total benefit of any doubt. And this indefinite charity only works through a simultaneous introjection from the animal that extends the human. The animal is like me because I change (morph, mutate) like the animal. It is the openness to sharing, to ex-*change*, that we share (not any *same* fixed powers of sensibility, reason, language, or whatever else). It is not a matter of *knowing* what it is like to *be x*, but feeling open to becoming *x*.

The newfound object of enquiry for philosophy, animals, not only mediates our understanding of the animal but also our image both of what counts as a proper philosophical object and even of what counts as proper philosophy (and so "non-philosophy") itself.[36] Philosophy is no longer unconditioned, free thought, but becomes highly conditionable. The reality effects of both cinema and philosophy are indeed Pavlovian in essence, because both converge on the animal from internal and external directions, centripetally and centrifugally. Their shared power points to the conditionable animal immanent within "us" (as currently defined).

And this affects politics too, being the idea *most* consistent with Alain Badiou's and Jacques Rancière's view that the political event is one that remaps what counts as the political and non-political as such, and so who counts as one amongst the *polis*, as my neighbour or companion – as a "who," not a "what." There are others that perceptibly help make my self, that are "like" what it is like to be me – mirror-others, other-humans. But many others may only emerge through an expanded alterity. They are Laruellean in the sense of being "non-"human actors, "strangers [who] insert themselves 'by force' into a community," in zones of affectivity and oppression-revolt – mutual regards which can be seen, have affects on each other, if only initially in horror (Laruelle, "The Generic" 240). They are animals who sense each other's joy and pain, and feel a "solidarity" with those affects.

The background revolts of cinema are forms of multiple philosomorphism that force us to attend to other morphisms, ones that change us, change what counts as philosophy, what counts as thought, what counts as the political, and what counts as human. The political underground is the background. What will the animal revolution look like? Horrifying cinema outlines it for us: its many stories, told through a play of sound and light, offer us glimpses of various future programmes. It tests theories by developing new logics of visibility and invisibility, of what it means to belong or be expelled, to think and decide, to be vulnerable and to suffer, to force oneself "into a community," to be deemed alive. And the audience – so-called "human-animals" at the cinema – participate in this too. Why, though, is this glimpse so abject to some and so desirable to others? Or, in the terms set by this essay, why is an emergence from the background horrific to some and wondrous to others? In *The Enigmatic Body*, Jean-Louis Schefer writes that "at bottom, the cinema is an abattoir" – a public spectacle of death and deformation (Schefer 120). And, as Michael Grant glosses this, it may be that a monster is "nothing other than a perpetual suffering of love and its animal lament" (Grant 130). Perhaps there is also some guilt at the core of the horrific, but therewith also the possibility of reparation alongside its threat.

notes

1 See Kennedy.

2 See Williams; Clover.

3 See Mullarkey, *Philosophy*.

4 See Thomas for the difference between "mode" and "genre" here.

5 For more on the philosophical background to this non-philosophical use of the term "background," see Mullarkey, "Spirit."

6 This is something that Derrida himself avows – "I who always feel turned toward death" (*H.C. for Life* 36).

7 See Midgley 349 n. 21.

8 See Irigaray 141.

9 Deleuze continues that this is true of the becoming of the animal as well, in its "becoming something else." But his oblique reference to "something else" is in clear contrast with the order of human becomings (from woman through animal to imperceptible) – animal becoming does not interest him in as much as it might participate in anything affirmative *for it*.

10 See Mullarkey, *Post-Continental Philosophy* 36–41.

11 The reference to the dying rat cited in the above quotation stems from the use of rat poison by the character of Lord Chandos in Hugo von Hofmannsthal's epistolary fable. See Lawlor.

12 In *A Thousand Plateaus* (Deleuze and Guattari 316) there is the example of the Stagemaker bird as a kind of artist ("the stagemaker practices *art brut*. Artists are stagemakers, even when they tear up their own posters"), but this is on account of the transversal and inhuman nature of art rather than the becoming-human of the bird per se.

13 For a full critique of Badiou's illegitimate exclusion of animal advocacy from his concept of the political event, at least as described in *Being and Event*, see Mullarkey, *Post-Continental Philosophy* 117–21.

14 Remember that we cannot sidestep this charge (of top-down class-representationalism being operative within an event) by saying that at least *all* involved were human, because that begs the question as to what/who counts as human (that is, it predefines the human *intensionally*).

15 See the essays collected in Daston and Mitman.

16 The "non-" in "non-philosophy" should be seen in terms of the "non-" in "non-Euclidean" geometries, being part of a "mutation" that locates philosophy as one instance in a larger set of theoretical forms – it is not its negation but a "*generalisation*" of philosophy. See Laruelle, *En tant qu'un* 8, 99ff.; idem, *Philosophie* 247.

17 Compare this: Indeed, panpsychist thought has often been accused of anthropomorphisation; of "retrojecting purely human mental traits into the non-human world" (Harman 212). That is, rather than stopping at the point of saying ice has a world – its own "psychic reality" (ibid.) – some panpsychists might go as far as to say the ice is dreaming of the time when it was water. Likewise, much panpsychist thinking seems to leave the hierarchy and ontological distinction of human and non-human intact: "Panpsychists typically see the human mind as a unique, highly-refined instance of some more universal concept. They argue that mind in, say, lower animals, plants, or rocks is neither as sophisticated nor as complex as that of human beings" (Skrbina). But the response to this is not to say that humans and stones or humans and ice are the same. It is to establish a horizontal continuum where human thought may be construed as "a more complicated variant of relations already found amidst atoms and stones," but

without assuming that more complicated necess-arily equals better or more important or valuable or deserving of greater rights (Harman 212).

18 We must recognise that such a "leap" has always already been made, that this leap is a retro-active construction. See Laruelle, *Non-philosophy Project* 92–93.

19 See Nietzsche 119.

20 It will be in recent Japanese horror where inan-imate matter, especially optical devices, takes on life, that the horror in and of cinema becomes most perspicuous.

21 See Mullarkey's *Philosophy* for an analysis of the double-edged nature of this specific attempt at a reality effect created through manipulating real death.

22 Coats cited in Cohen 24 n. 39.

23 See Rossell.

24 On weird horror, see Miéville.

25 If there is an appearance of essentialism as regards both the meaning of horror and the animal for this essay too, one must keep in mind the processual and perspective "nature" of the two that is being forwarded; that is, these "revolting" animals only indicate an abyss – wherein *everything* lives (the "animal" here really being a place-holder or Trojan-horse for animism) – to an anthropocentric point of view, which is itself always shifting. (Though, of course, these shifts in what counts as *anthropos* may be rationalised as the discovery of the *same* human essence in new populations, rather than a mutation of that essence – which is really to say that it has no essence.) Hence, the revolting animal is both "hor-rific" in Carroll's terms (repulsively interstitial) *and* political (because it redraws the lines around what/who count as living beings). But those lines are always perspectival. Latour (137) captures this thought in the following:

> The expression "anthropomorphic" con-siderably underestimates our humanity. We should be talking about morphism. Morphism is the place where technomorphisms, zoo-morphisms, phusimorphisms, ideomorph-isms, theomorphisms, sociomorphisms, psychomorphisms, all come together. Their alliances and their exchanges, taken together, are what define the *anthropos*. A weaver of

morphisms – isn't that enough of a definition? The closer the *anthropos* comes to this distri-bution, the more human it is. The farther away it moves, the more it takes on multiple forms in which its humanity quickly becomes indiscernible [...]

It is this loss of discernibility that is horrific to the self-styled "humans." Matt Hills' event-based defi-nition of art-horror looks at *Blair Witch*, *The Haunt-ing* and *Event Horizon* in terms of which they all fail to conform to a "standard cinematic form" (to use C.S. Tashiro's term) built on "middle distances, the cutting patterns of spaces calculated to human scale" (Hills 146ff.).

26 It has even been proposed that a "Japanese National Science" could "reconcile us with nature instead of opposing it." Here we see the possibility of an identification of the human with the inert that would be less likely to bring with it the usual con-notations of a horrifying *reduction* in value: see Stengers 133.

27 See Powell 27.

28 There are apparently possessed photographs in Western horror films like *The Omen* or *Final Desti-nation 3*, *The Shining* and *The Others*, as well as seemingly demonic televisions in *Poltergeist* and mirrors in *Mirrors*, but these are usually signs of, or portals to, other realms of life, rather than living entities in themselves.

29 And again:

> the film thus associates ubiquitous techno-logical mediation – that is, the cameras, television sets, videocassette recorders, tele-phones and other such hardware fore-grounded throughout the film – with the intrusion of "posthuman" otherness into contemporary cultural life. (41)

30 As Dennis Giles writes (37): "cinema is never the raw vision of desire [...] the experience of cinema is simultaneously a screening and a screen-ing off."

31 See Mullarkey, *Reverse Mutations*.

32 Cited in Badiou, *Infinite Thought* 127.

33 See Kolozova.

34 The following all-too-condensed discussion of Laruelle's work can be supplemented with readings

contained in Mullarkey and Smith which aim to explain as best as possible the challenging and novel aspects of Laruelle's very "non-standard" philosophy.

35 See Laruelle, *Struggle and Utopia*.

36 Eugene Thacker also links this non-philosophy to a horror of the non-human world "[...] *'horror' is a non-philosophical attempt to think about the world-without-us philosophically*" (*In the Dust of this Planet* 9). And further:

> what genre horror does do is it takes aim at the presuppositions of philosophical inquiry – that the world is always the world-for-us – and makes of those blind spots its central concern, expressing them not in abstract concepts but in a whole bestiary of imposs-ible life forms – mists, oozes, blobs, slime, clouds, and muck. Or, as Plato once put it, "hair, mud, and dirt."

bibliography

Agamben, Giorgio. *The Open: Man and Animal*. Trans. Kevin Attell. Palo Alto: Stanford UP, 2004. Print.

Asma, Stephen T. *On Monsters: An Unnatural History of our Worst Fears*. Oxford: Oxford UP, 2009. Print.

Badiou, Alain. *Being and Event*. Trans. Oliver Feltham. London: Continuum, 2006. Print.

Badiou, Alain. *Ethics: An Essay on the Understanding of Evil*. Trans. Peter Hallward. London: Verso, 2001. Print.

Badiou, Alain. *Infinite Thought: Truth and the Return to Philosophy*. Trans. and ed. Oliver Feltham and Justin Clemens. London: Continuum, 2003. Print.

Badiou, Alain. *Logics of Worlds*. Trans. Alberto Toscano. London: Continuum, 2009. Print.

Badiou, Alain. "Review of Gilles Deleuze, *The Fold: Leibniz and the Baroque*." *Gilles Deleuze and the Theatre of Philosophy*. Ed. Constantin V. Boundas and Dorothea Olkowski. London: Routledge, 1994. 51–69. Print.

Badiou, Alain. *Le Siècle*. Paris: Seuil, 2005. Print.

Badley, Linda. *Film, Horror and the Body Fantastic*. Westport: Greenwood, 1995. Print.

Balmain, Colette. *Introduction to Japanese Horror Film*. Edinburgh: Edinburgh UP, 2008. Print.

Burt, Jonathan. *Animals in Film*. London: Reaktion, 2002. Print.

Burt, Jonathan. "Morbidity and Vitalism: Derrida, Bergson, Deleuze, and Animal Film Imagery." *Configurations* 14 (2006): 157–79. Print.

Carroll, Noël. "Film, Emotion, and Genre." *Passionate Views: Film Cognition, and Emotion*. Ed. Carl Plantinga and Greg Smith. Baltimore: Johns Hopkins UP, 1999. 21–47. Print.

Carroll, Noël. "Why Horror?" *Horror Film Reader*. Ed. Mark Jancovich. London: Routledge, 2002. 33–45. Print.

Cholodenko, Alan. "The Animation of Cinema." *Semiotic Review of Books* 18.2 (2008): 1–10. Print.

Clover, Carol J. "Her Body Himself: Gender in the Slasher Film." *Representations* 20 (1987): 187–228. Print.

Cohen, Jerome J. *Monster Theory*. Minneapolis: U of Minnesota P, 1996. Print.

Daston, Lorraine, and Gregg Mitman, eds. *Thinking with Animals: New Perspectives on Anthropomorphism*. New York: Columbia UP, 2006. Print.

Deleuze, Gilles. *Cinema 1: The Movement-Image*. Trans. Hugh Tomlinson and Barbara Habberjam. London: Athlone, 1986. Print.

Deleuze, Gilles, and Félix Guattari. *Anti-Oedipus*. Trans. Robert Hurley, Mark Seem, and Helen R. Lane. London: Athlone, 1984. Print.

Deleuze, Gilles, and Félix Guattari. *A Thousand Plateaus*. Trans. Brian Massumi. London: Athlone, 1987. Print.

Deleuze, Gilles, and Félix Guattari. *What is Philosophy?* Trans. Hugh Tomlinson and Graham Burchill. London: Verso, 1994. Print.

Derrida, Jacques. *The Animal that Therefore I Am*. Trans. David Wills. New York: Fordham UP, 2008. Print.

Derrida, Jacques. "'Eating Well', or the Calculation of the Subject: An Interview with Jacques Derrida." *Who Comes after the Subject?* Ed. Eduardo Cadava, Peter Connor, and Jean Luc-Nancy. Trans. Peter Connor and Avital Ronnell. London: Routledge, 1991. 96–119. Print.

Derrida, Jacques. *H.C. for Life, That is to Say*. Trans. Laurent Milesi and Stefan Herbrechter. Palo Alto: Stanford UP, 2006. Print.

Garcia, Tristan. *Forme et objet. Un traité des choses.* Paris: PUF, 2011. Print.

Giles, Dennis. "Conditions of Pleasure in Horror Cinema." *Planks of Reason: Essays on the Horror Film.* Ed. Barry Keith Grant and Christopher Sharret. Lanham, MD: Scarecrow, 2004. 36–49. Print.

Grant, Michael. "On the Question of the Horror Film." *Dark Thoughts: Philosophical Reflections on Cinematic Horror.* Ed. Stephen Jay Schneider and Daniel Shaw. Lanham, MD: Scarecrow. 120–37. Print.

Haraway, Donna. *When Species Meet.* Minneapolis: U of Minnesota P, 2007. Print.

Harman, Graham. *Prince of Networks: Bruno Latour and Metaphysics.* Melbourne: Re:press, 2009. Print.

Hills, Matt. *The Pleasures of Horror.* London: Continuum, 2005. Print.

Irigaray, Luce. *This Sex Which is Not One.* Trans. Catherine Porter with Carolyn Burke. Ithaca, NY: Cornell UP, 1985. Print.

Kennedy, John S. *The New Anthropomorphism.* Cambridge: Cambridge UP, 1992. Print.

Kolozova, Katerina. *The Lived Revolution: Solidarity with the Body in Pain as the New Political Universal.* Skopje: Evro-Balkan, 2010. Print.

Laruelle, François. "The Generic as Predicate and Constant: Non-philosophy and Materialism." *The Speculative Turn: Continental Materialism and Realism.* Ed. Levi Bryant, Nick Srnicek, and Graham Harman. Melbourne: Re:press, 2011. 237–60. Print.

Laruelle, François. *The Non-philosophy Project: Essays by François Laruelle.* Ed. Gabriel Alkon and Boris Gunjevic. New York: Telos, 2012. Print.

Laruelle, François. *Philosophie et non-philosophie.* Liège: Mardaga, 1989. Print.

Laruelle, François. *Struggle and Utopia at the End Times of Philosophy.* Trans. Drew S. Burke and Anthony Paul Smith. Minneapolis: Univocal, 2012. Print.

Laruelle, François. *En tant qu'un. La Non-philosophie expliquée au philosophes.* Paris: Aubier, 1991. Print.

Latour, Bruno. *We Have Never Been Modern.* Trans. Catherine Porter. Cambridge, MA: Harvard UP, 1993. Print.

Lawlor, Leonard. "Following the Rats: Becoming Animal in Deleuze and Guattari." *SubStance* 37.3 (2008): 169–87. Print.

Lippit, Akira. "The Death of the Animal." *Film Quarterly* 56 (2002): 9–22. Print.

Lippit, Akira. *Electric Animal.* Minneapolis: U of Minnesota P, 2000. Print.

McRoy, Jay. "Case Study: Cinematic Hybridity in Shimizu Takashi's *Ju-On: The Grudge.*" *Japanese Horror Cinema.* Ed. Jay McRoy. Edinburgh: Edinburgh UP, 2005. Print.

Merleau-Ponty, Maurice. *The Phenomenology of Perception.* Trans. Colin Smith. London: Routledge, 1962. Print.

Midgley, Mary. *Beast and Man: The Roots of Human Nature.* London: Routledge, 1978. Print.

Miéville, China. "M.R. James and the Quantum Vampire: Weird; Hauntological: Versus and/or and and/or or?" *Collapse IV* (2008): 105–28. Print.

Mullarkey, John. *Philosophy and the Moving Image: Refractions of Reality.* Basingstoke: Palgrave Macmillan, 2010. Print.

Mullarkey, John. *Post-Continental Philosophy: An Outline.* London: Continuum, 2006. Print.

Mullarkey, John. *Reverse Mutations: Laruelle and Non-human Philosophy.* Forthcoming. Print.

Mullarkey, John. "Spirit in the Materialist World: Revisionary Metaphysics and the Horrors of Philosophy." *Angelaki.* Spec. ed. on *Immanent Materialisms: Speculation and Critique.* Ed. Charlie Blake and Patrice Haynes. Forthcoming. Print.

Mullarkey, John, and Anthony Paul Smith, eds. *Laruelle and Non-philosophy.* Edinburgh: Edinburgh UP, 2012. Print.

Nietzsche, Friedrich. *On the Genealogy of Morals and Ecce Homo.* Trans. Walter Kaufmann and R.J. Hollingdale. New York: Vintage, 1989. Print.

Nowak, Lars. "Mother's Little Nightmare: Photographic and Monstrous Genealogies in David Lynch's *The Elephant Man.*" *Journal of European Popular Culture* 1.1 (2010): 69–83. Print.

Oliver, Kelly. *Animal Lessons: How they Teach us to be Human.* New York: Columbia UP, 2009. Print.

Powell, Anna. *Deleuze and Horror Film*. Edinburgh: Edinburgh UP, 2007. Print.

Rossell, Deac. *Living Pictures: The Origins of the Movies*. New York: State U of New York P, 1998. Print.

Schefer, Jean-Louis. *The Enigmatic Body: Essays on the Arts*. Cambridge: Cambridge UP, 1995. Print.

Shaviro, Steven. *The Cinematic Body*. Minneapolis: U of Minnesota P, 1993. Print.

Skrbina, David. "Panpsychism." *Internet Encyclopedia of Philosophy*. 2007. Web. 15 Jan. 2013. <http://www.iep.utm.edu/panpsych>.

Stengers, Isabelle. *Power and Invention*. Trans. Paul Bains. Minneapolis: U of Minnesota P, 1997. Print.

Thacker, Eugene. *In the Dust of this Planet: Horror of Philosophy Vol. 1*. Ropley: Zer0, 2011. Print.

Thacker, Eugene. "Thought Creatures." *Theory, Culture and Society* 24 (2007): 327–29. Print.

Thomas, Deborah. *Beyond Genre: Melodrama, Comedy and Romance in Hollywood Films*. Moffat: Cameron, 2000. Print.

Wells, Paul. *The Animated Bestiary: Animals, Cartoons, and Culture*. Chapel Hill, NC: Rutgers UP, 2009. Print.

Wells, Paul. *The Horror Genre: From Beelzebub to Blair Witch*. London: Wallflower, 2000. Print.

Williams, Linda. "Film Bodies: Gender, Genre, and Excess." *Film Quarterly* 44.1 (1991): 2–13. Print.

Agency is, I believe, distributed across a mosaic, but it is also possible to say something about the kind of striving that may be exercised by a human within the assemblage. This exertion is perhaps best understood on the model of riding a bicycle on a gravel road. One can throw one's weight this way or that, inflect the bike in one direction or toward one trajectory of motion. But the rider is but one actant operative in the moving whole.

    Jane Bennett, Vibrant Matter

How do you know but ev'ry bird that cuts the airy way
Is an immense world of delight, closed by your senses five?
    William Blake, "Marriage of Heaven and Hell"

At the 2009 conference of the Society for Literature, Science and the Arts (SLSA), I delivered a paper designed to investigate the meaning of the shift from the cyborg to the animal in posthumanist philosophy. This shift is evident in the archives of SLSA panel titles, and is registered most palpably in Donna Haraway's abandonment of cyborg theory for a tongue-kissing session with her dog. I opened the talk with a reprogramming of this scene from Haraway's *When Species Meet*, a poem of sorts designed to provoke fellow delegates on an animal studies panel. Ultimately my ruse was to ask: "Why stop at the animal? Why not consider insects, trees, even rocks as objects of posthumanist speculation?" W.J.T. Mitchell asked this very question in his Foreword to Cary Wolfe's *Animal Rites*. What both Mitchell and I had failed to realize is that this posthumanist shift from cyborg to animal

marcel o'gorman

SPECULATIVE REALISM IN CHAINS
a love story

to "things" was already well underway; it was underscored at the conference by a series of panels on Whitehead and a keynote address by Ian Bogost on "Alien Phenomenology." Bogost's radical speculation delivered in earnest what I was attempting to deliver as rhetorical irony. My attempted provocation, then, was not radical at all. Still, I would like to revisit my de-ironized text in order to rescue its grounding concept, which is that the trajectory of posthumanist philosophy would ultimately land in the world of things, and this would be driven by: (a) a desire to *connect* with the nonhuman world in ways that ignore the lessons of post-structuralism; and (b) a romantic wonder about the infinite world of things acting in the universe.

For indeed, romance is what drives this attraction to the cyborg, the animal, and more recently, the inorganic thing. And by "romance" I mean not only the literary genre – I am unabashedly talking about love. Above all, this is an essay about posthumanist love.

requiem for the cyborg

Mikado continues to colonize all my cells – a sure case of what the biologist Lynn Margulis calls symbiogenesis. I bet if you were to check our DNA, you'd find some potent transfections between us. We inhabit not just different genera and divergent families but altogether different orders. Yet I can't resist the feel of him nestling deep in my sensitive backside. Together, generating galvanic friction, we've worn out the crotch of many a pair of black Levi's.

I've had to tether him outside my lab, where he waited patiently as I taught a class in such worn jeans, pantlegs stained with dirty lubricant. I've also had to leave him in suspense, behind the locked doors of an old shed, vulnerable to the dilations and contractions of a Canadian winter. These thoughts trouble me.

How would we sort things out? Biped, bipedal; working man, workhorse; saddle-weary rider, seasoned courser. One has a photo ID Ontario Driver's license; the other has a serial number etched on his underside. One has an Irish name in spite of his French descent; the other has a Japanese name but cannot trace his origins beyond Canada.

One of us, product of a complex organic assemblage, is called "white male." The other, product of an intricate inorganic assemblage, is called "bicycle." Each of these names designates a different racial and gendered discourse, and we both inherit their consequences in our bodies.

One of us is too old for adventure racing, but denies it in spite of himself; the other is visibly rusty, but still runs like a well-oiled machine. And we race together on rough trails at the county landfill where Mikado will surely be laid to rest some day, each outing bringing him closer to this demise.

I'm sure our genomes are more alike than they should be. Some molecular record of our touch in the codes of living will surely leave traces in the world, no matter that we are each reproductively silenced males, both fixies.

We have given each other scars, worn each other down. Mikado has intruded deeply into my flesh, with all its eager immune system receptors. Who knows where my chemical receptors carried his messages or what he took from my cellular system for distinguishing self from other and binding outside to inside?

Mikado is quick to shriek and squeal if I push too hard, and has thrown me to the ground in response to poor handling. We have had forbidden conversation; we have had rectal intercourse; we are bound in telling story on story with nothing but the facts.

We are training each other in acts of communication we barely understand. We are, constitutively, companion species. We make each other up, body to body, broken-down frame to broken-down frame. Significantly other to each other, in specific difference, we signify through our bodies a nasty developmental infection called love. This love is a historical aberration and a naturalcultural legacy.

This plagiaristic reworking of Haraway's text marks a turn from the animal to the thing in posthumanist speculation. In short, anything Haraway can say about her dogs, I can say about my bike, with no less care or concern, no less love. We might label this passage as fetishistic, vitalistic, animistic, or anthropomorphic. It is perhaps all of these things, but above all it is an act of empathetic imagination, a generative text designed to provoke thought about the relationship between humans and things, or, more radically, about humans *as* things. The text is therefore an "evocative object," a concept I will discuss in greater detail below. For the moment, let's stay seated on the bicycle, ever chased by Haraway's dog, and consider what it means to shift gears from animal to inorganic thing in posthumanist thought.

In *When Species Meet*, Haraway notes the common misconception that "when people hear the term *companion species*, they tend to start thinking about 'companion animals,' such as dogs, cats, horses, miniature donkeys, tropical fish, fancy bunnies, dying baby turtles, ant farms, parrots, tarantulas in harness and potbellied pigs" (17). This list, despite its heterogeneity, is too delimiting in Haraway's view, which posits that the term "companion species" should be considered less "a category than a pointer to an ongoing 'becoming with,' [...] a much richer web to inhabit than any of the posthumanisms on display after (or in reference to) the ever-deferred demise of man" (17). It takes very little imagination, then, to add to Haraway's list of companions a bicycle. This shift in gears toward the inorganic, yet another strategy for decentering the *speciesist* human subject, is evident in the recent growth of speculative realist (SR) philosophies and object-oriented ontology (OOO). But the shift could not be accomplished without love, a love that is apparent both in the recital of a list ("ant farms, parrots, tarantulas") and in the concept of "becoming-with."

For the collector of things, the list is an object of love. It is at once an indicator of finitude (here's what I have) and infinity (here's what I desire). Just as importantly, it is an indicator of possession (this is what I own, this is what I know). Ian Bogost has coined the term *Latour litany*, inspired by Bruno Latour's "parliament of things," to describe this tendency toward listing in posthumanist philosophy. The list, Bogost suggests in his blog, "underscores the rich diversity of things" in a way that cannot be accomplished with ordinary narrative logic and syntax. Bogost owes his own list-obsession more to Graham Harman than to Bruno Latour. Harman's litanies are a key rhetorical device in his OOO, as evidenced in the following passage, which sums up the entire thesis of the book *Tool Being*:

> inanimate objects are not just manipulable clods of matter, not philosophical deadweight best left to "positive science." Instead, they are more like undiscovered planets, stony or gaseous worlds which ontology is now obliged to colonize with a full array of probes and seismic instruments – most of

them not yet invented [...] Instead of aloof reflections on the enframing mechanisms of technology, it ought to be possible to discuss subways and radio telescopes. And enough with clever references to the "dice throw" as an avant-garde literary image: dice and slot machines and playing cards *themselves* should be our theme – as should fireworks, grasshoppers, moonbeams, and wood. (19–20)

What the list achieves for both Harman and Bogost is the micro representation of an infinity of things jostling up against one another, "rubbing shoulders," as Bogost puts it, in a fractal dance of becoming: "It happens fast and hot, the universes of things bumping and rubbing against one another in succession, chaining together like polymers" (*Alien Phenomenology* 25). The overt eroticism of this statement reveals the truth: in the final analysis, OOO is a "fast and hot" love story.

To Bogost's chaining of polymers we might link Jane Bennett's use of the term "assemblage," borrowed from Deleuze and Guattari. The concept of assemblage challenges notions of phenomenological verticality or speciesism, permitting Bennett to "rattle the adamantine chain that has bound materiality to inert substance and that has placed the organic across a chasm from the inorganic" (57). Bennett's avowal of love rings clear in her commitment to an ethics of caring concern, which unfolds in her assemblages of human and nonhuman actants, assemblages that call for increased attention to and responsibility for the materiality of being:

> I am a material configuration, the pigeons in the park are material compositions, the viruses, parasites, and heavy metals in my flesh and in pigeon flesh are materialities, as are neurochemicals, hurricane winds, E. coli, and the dust on the floor. Materiality is a rubric that tends to horizontalize the relations between humans, biota, and abiota. It draws human attention sideways, away from an ontologically ranked Great Chain of Being and toward a great appreciation of the complex entanglements of humans and nonhumans. (111–12)

Whereas Ian Bogost's lists allow him to luxuriate in speculative ruminations on the infinite modalities of being, Bennett's horizontalizing ontology is a vehicle for ecocritical care. In spite of these rather different uses for OOO, both are invocations of love. In response to Bennett's romantic confession of an "irrational love of matter" (61), we have Ian Bogost's orgiastic pronouncement that "anything is thing enough to party" (*Alien Phenomenology* 23). But neither of these invocations confesses with honesty to the fast and hot eroticism that is at play in the groupings and becomings of speculative realism. It's time to get seriously horizontal.

Haraway's make-out scene with her dog is emblematic of the erotic thrust of posthumanist philosophy and its desire to make contact with nonhuman others. This desire, as Bogost demonstrates, is being pursued to pornographic extremes in a contest of unchained empathy that assigns agency to increasingly exotic or "alien" nonhuman entities. At its worst, this contest reeks of an anthropocentric colonization, albeit unwitting, as in Graham Harman's will to "colonize" inanimate objects "with a full array of probes and seismic instruments" (19). Harman's choice of words speaks for itself here, revealing a colonial and phallogocentric drive that threatens to undermine the otherwise posthumanist, post-gender, post-subject bent of OOO. In this boys' club context, Bogost's term "unit operations" takes on a whole new phallic meaning. At times, the *unit* seems to be running the show for a philosophy of sublimated desire.

At the other end of the spectrum, marking the obscene end of OOO well before Bogost's intervention, is Mario Perniola, who describes his erotic ontological aphorisms as a form of "speculative extremism." Perniola, lifting the veil on the erotics of speculative realism, pushes to obscenity the desire for contact with the inorganic (1). Inspired by Walter Benjamin's description of the anthropomorphic drawings of plants and clothes rendered by J.J. Grandville, Perniola's speculation takes root in the "sex appeal of the inorganic" (3). As Benjamin suggests in "Paris, Capital of the

Nineteenth Century," by pushing fashion "to its extremes" Grandville "revealed its nature," and that nature involves human desire, inorganic matter, sex, and death. Fashion, in Grandville's cosmology, "prostitutes the living body to the inorganic world. In relation to the living it represents the rights of the corpse. Fetishism, which succumbs to the sex appeal of the inorganic, is its vital nerve" (79). And yet it is essential not to confuse fetishism, via Marx and Freud, with the sex appeal of the inorganic. Perniola is not after a Marxist critique of speculative phenomenology (a project that would certainly be worthy of pursuing); rather, by forcing such speculation to its extreme, he reveals its sensual and especially erotic underside. All that rubbing together of "the stuff of being" (Bogost, *Alien Phenomenology* 27) documented in the work of Harman and Bogost, and those vibrant "entanglements" (Bennett 112) of organic and inorganic matter in the work of Bennett and Bryant reveal a universe of becoming determined by the sense of touch, or put more plainly, a "world of things that feel" mobilized by the sex appeal of the inorganic (44).

Perniola's positioning of sex appeal at the core of a speculative exercise, an exercise that is less pornographic than it is *pornohaptic*, is perhaps best explained on his own terms, quoted here at length so that the reader can witness the erotic foreplay in Perniola's rhetoric:

> Lie down in a state of absolute rest with your eyes closed as if you were dead. You are deaf, mute and blind and you remain like this despite any appeal, incitement or solicitation. Of all the senses only the sense of touch is left but you cannot exercise it actively. All your attention is concentrated on what brushes against you, touches, you, feels you and only on the basis of this pressure are you able to picture the form and the shape of the hands that touch you and caress you, that penetrate into the folds and into the cavities of your flesh. You have never seen, or heard anything, or know the meaning of the silent question of what plods panting over you. Also, all reaction or reflex is precluded

to you. You must not give any sign of life, or laugh if you are tickled, or utter moans, cries, or react minimally to any stimulation that becomes progressively more intrusive, that persists in provoking a reaction, that does not withdraw but attacks those parts that they imagine are the most sensitive. (34)

What might otherwise look or feel like a masochistic act of submission is not, however, tied to any sort of violence or domination. Perniola's inorganic things, which include human things, are not "manipulable clods of matter" (Harman 19); rather, the relationship between things here is a matter of "interpenetration," of haptic relationality within the "unlimited space opened up by the disappearance of the subject" (44). As Perniola notes, "this faked death [...] is without frigidity. Your abandonment does not exclude: on the contrary it implies that whoever devotes himself to you will move your legs, open your mouth, lift your head if he wishes" (34). The sex appeal of the organic lays bare the horizontality of posthumanist thinking, which is ultimately a discourse of openness to promiscuous intercourse, linking, joining together – a discourse of copulation.

Discourse – more specifically, philosophical discourse – is truly what is at stake here. Recalling Perniola's image of the thing in submission, it is essential to note that "flat ontology," as Manuel DeLanda and Levi Bryant have called it, requires a certain suspension of subjectivity, an end to what Perniola calls the "orgasmomania" of contemporary philosophy. But this does not mean an end to sex, an abstention or ascesis, which is how one might describe the philosophy of Heidegger, whose prophylactic logic fails to shelter him from the sex appeal of the inorganic. According to speculative realism, try as you might to resist, there is no choice but to rub and jostle against other things in the infinite, fractal, horizontal orgy of being. What Perniola suggests, without reserve, is that the orgy is reserved for philosophers, for those who are able to fully give themselves over to the inorganic. It is here, in his description of "philosophical-sexual cyborg,"

that Perniola pushes speculative realism to its breaking point:

> When you find the realization of the Cartesian thing that feels in the cunnilingus or in the fellatio of your partner, when you notice in the coherent and rigorous unfolding of philosophic prose the inexorable movement that brings you to lick the cunt, the cock or the arse of your partner who has become a neutral and limitless extension of cloth variously folded, when you yourself are able to offer your body as a desert or a heath so that it can be traversed by the detached and inexorable examination of the eye, the hands and the mouth of your lover, when nothing else interests you or excites you or attracts you besides repeating every night the ritual of the double metamorphosis of philosophy into sex and sex into philosophy, then, maybe, [...] you have celebrated the triumph of the thing over everything, you have led the mind and the body to the extreme regions of the non-living, where, perhaps, they were always already directed. (16)

Here, speculative realism is laid bare, revealed not as an ontology at all but as philosophical calisthenics, a sort of Kegel exercise designed to improve the quality of human intercourse with things. In this sexual-philosophical encounter, a radical objectification of the human is chained to an equally radical subjectification that can only be manifested in and through philosophy. What Perniola's scene of intercourse reveals is that, ultimately, extreme speculation of the brand described in this essay may not be rooted in physical eros at all but in language, in a philosophical intercourse akin to Socrates' understanding of love in *Phaedrus*. Above all, this scene of copulation between humans freed of subjectivity, this "metamorphosis of philosophy into sex and sex into philosophy," reveals a veiled desire for unity and holism, for a perfectly transparent communication, a philosophical interpenetration that, properly understood, is more about love than it is about sex.

It is time to confess that, thus far, I have been rather unrigorous with my use of the term "love." I have identified a potential erotic under-belly in the work of speculative realist thinkers. But eros is not the only type of love that characterizes the work of SR and OOO; it is simply the most obvious manifestation of a love for objects that might also be characterized as both *philia* (love by filiation, brotherly love serving mutual goals) and *agape* (self-effacing, Christian love, selfless love such as that professed in the gospels). As I noted in the first paragraph of this essay, in his introduction to Cary Wolfe's book *Animal Rites*, W.J.T. Mitchell unwittingly gestures toward the OOO phenomenon in a statement designed to be a provocative exaggeration:

> Let us suppose, finally, that all these issues have been worked out and the rights of all animals, high and low, have been established. Would that be the end? Or would it then be time to turn to the rights of fruits and vegetables? Erasmus Darwin noted long ago that "the loves of plants" are essential to their lives. Does that give them a claim to some sort of rights? (xi)

What Mitchell is attempting to do here is distance Wolfe's philosophical project from the "self-indulgent breast-beating that encourages moralistic, sentimental posturing while doing nothing to change the lot of animals" (x). Animals have the potential, in Mitchell's mind, to be the "latest candidates in an endless procession of victims – women, minorities, the poor – clamoring for rights and justice, or just a modicum of decent treatment" (x). To illustrate how this procession might extend to plants and then inanimate objects, Mitchell quotes William Blake's anthropomorphic poem "Ah! Sunflower." Had Mitchell quoted Blake's "The Clod and the Pebble," he would have demonstrated greater foresight about the philosophical turn toward objects that was to follow animal studies. For this is a turn rooted in love. With that, I quote Blake's poem here in full to further develop the love story that is OOO:

"Love seeketh not itself to please,
Nor for itself hath any care,

But for another gives its ease,
And builds a heaven in hell's despair."

So sung a little Clod of Clay,
Trodden with the cattle's feet,
But a Pebble of the brook
Warbled out these metres meet:

"Love seeketh only Self to please,
To bind another to its delight,
Joys in another's loss of ease,
And builds a hell in heaven's despite."

In this poem of contraries, typical of Blake's *Songs of Innocence and Experience*, we see divergent perspectives on love: one innocent and the other experienced to the point of cynicism. The first love illustrates the very well-known concept of love as *agape*, selfless Christian love. The second love, on the other hand, might be identified as *philia*, which in *Nicomachean Ethics* (Book VIII), Aristotle describes in terms of a business transaction, a relationship of utilitarian, mutual enjoyment. Perhaps both of these loves are present in the work of Bennett, Bryant, Harman, Bogost and others. Bennett's vitalist, ecocritical materialism seems motivated by a Christian ethics, a version of "love thy neighbor as thyself" that includes as neighbors pigeons, hurricane winds, and *E. coli*, among other vibrant things. Bogost's love, on the other hand (and we should remember Perniola here as well), which revels in the "chaining together" of tiny universes, seems to "bind another to its own delight" for the sake of generating a radical ontology. Mitchell could not have predicted that even Blake, in his conspicuous anthropomorphism, could not match the extremes to which OOO would go in its proclamation of philial love for objects. If you were to throw Blake's clod into the warbling brook and watch it disperse into a myriad individuated particles, then you would have a better sense of the love-sickness that is OOO.

The term "love-sickness" brings to mind Socrates' discourse on love in *Phaedrus*, as mentioned briefly above. Of central interest here is Socrates' description of the four forms of madness, the fourth being mad love, which he defends as a divine state of being, the lover's glimpse of a once-unified body and soul manifested in the beautiful object of love. In his defense of the mad lover, Socrates introduces the term *anteros*, a counter-eros that binds philia and eros together in a mutual and self-reflexive eroticism:

> And thus he [the mad lover] loves, but he knows not what; he does not understand and cannot explain his own state; he appears to have caught the infection of blindness from another; the lover is his mirror in whom he is beholding himself, but he is not aware of this. When he is with the lover, both cease from their pain, but when he is away then he longs as he is longed for, and has love's image, love for love (*Anteros*) lodging in his breast, which he calls and deems not love but friendship only, and his desire is as the desire of the other, but weaker; he wants to see him, touch him, kiss, embrace him, and not long afterwards his desire is accomplished. (117)

The mutual love of *anteros* emerges when both lovers recognize beauty – in its Platonic, prelapsarian ineffability – in one another. As John Durham Peters suggests in *Speaking into the Air*, "Sexual desire thus is not demeaned as base by Socrates but considered an intimation of cosmic homesickness" (44). There is perhaps some of this cosmic homesickness in the work of OOO and SR, and I will suggest that it allows us to chain together Plato, the Romantics, and OOO, which are all bound by a desire for holism, unity, and the overcoming of impossible barriers to communication.

Peters' wide-ranging examination of communication media and human desire takes him from Plato's dialogues, to the gospels of the New Testament, to nineteenth-century mystics, and finally to contemporary communications media. All of these attempts at communication, he suggests, from Plato's dialogues to SETI's extraterrestrial hailing, are emblematic of the human desire to overcome the "gap between sending and receiving" (151). In spite of what might be described as Peters' uncareful dismissal of poststructuralism, he provides another way of understanding OOO as an attempt to regain a lost unity, a

wholeness that it has been the work of poststructuralism to unravel and debunk. The work of both psychics and SETI, Peters suggests, deals

> with the most poignant human concerns: mourning, cosmic loneliness, contact with the dead and distant (psychical research) or alien and distant (SETI). Both are moved by faith in the other's existence without the ability to take hold of a sure connection. Both imagine a universe humming with conversations we are unable, for whatever reasons, to tap. (248)

The more alien and distant the other we attempt to contact, the greater the act of love, the greater the potential for proving a cosmic unity that binds all things. In Peters' words, "empathy with the inhuman is the moral and aesthetic lesson that might replace our urgent longing for communication" (248). Seen in this light, Bogost's *alien phenomenology*, within which even the Roswell alien might "rear its head," can be seen as a philosophy of empathetic imagination, of love, driven by cosmic loneliness and a desire to be recognized by and connect with a seemingly inaccessible universe.

As I have alluded to in a previous pun, this desire can be described as romantic in both the literary and amorous sense of the word, and it is thus nothing new or provocative. As Julia Martin argues, Blake's philosophical cosmology, in which "every particle of dust breathes forth its joy" (*Europe*, lines 18–19), requires "acts of imaginative identification that are involved in sympathy or love or compassion" (60). Echoing Jane Bennett, Martin suggests that the romantic identification with things as exemplified in Blake's work emerges from a view of the cosmos as an "interdependent network of care," which calls for awareness about ecological issues, in all their political and philosophical complexity (53). It does not take a suspension of disbelief to see a trace of romanticism in OOO. This is most prevalent perhaps in the work of Bogost, whose desire to flee "from the dank halls of the mind's prison toward the grassy meadows of the material world" is conspicuously reminiscent of Blake's desire in *Marriage of Heaven and Hell* to open the "doors of perception" (Bogost, *Alien Phenomenology* 39; Blake 39). Blake's goal was to flee the *Single Vision* which he, like Bogost, "identified in the reductionist gaze of eighteenth-century rationalism" (62). Bogost could easily adopt Los's motto from Blake's poem *Jerusalem*: "I must Create a System or be enslav'd by another Mans" (153).

It may be more accurate to view Bogost's work not as a romantic ideology but more specifically as *technoromantic*, following in the footsteps of such writers as Michael Heim and Howard Rheingold. In "Cyberspace and Heidegger's Pragmatics," Richard Coyne identifies a distinctly romantic ethos in rhetorics about technoculture, one that would lead him to coin the term *technoromanticism*:

> The dominant ethos is now romanticism: a focus on subjectivity, a new metaphysics of proximity, a revival of the early socialist dream of community, a disdain for the constraints imposed by the body, embracing the holistic unitary patterning of chaos theory, the representation of the object world, a hope for its ultimate transcendence through the technologies of cyberspace, and a quest for a better, fairer more democratic future. (349)

Object-oriented ontology unquestionably shares many, if not all, of these characteristics, except, of course, that its hope for transcendence derives not from pondering the infinity of cyberspace but from pondering the infinite possibilities of object being. It is fitting, then, and not at all random or coincidental, that the term "object-oriented," as Bogost reminds us, comes from computer science. Just as OOO provides an outlet for mourning the lost unity of language after poststructuralism, it also provides an opportunity to assume a technoromantic ethos after the debunking of emancipatory cybernarratives, from Heim's erotics of cyberspace to the idealistic claims of George P. Landow and other hypertext theorists of the 1990s. I will leave this evocative genealogy from cyberrhetoric to OOO for another essay. To avoid straying too far from love, I would like to focus on yet another tradition from

which OOO emanates, further demonstrating its status as a philosophy for lovers.

Given the romantic tendencies of OOO and SR, it is no surprise that *Alien Phenomenology* ends with an essay on the importance of injecting *wonder* into contemporary philosophy. Citing Graham Harman, Bogost writes that "wonder is a sort of *allure* that real objects use to call at one another through enticement and absorption [...] Wonder is a way objects orient" (125). This is a philosophy shared by many in the field of ecocriticism, which unabashedly accepts its genealogy of romantic thought. As David Sander suggests, "it is precisely in the moment of wonder that the imagination is challenged to understand and apprehend the environment, opening a space in literature for the negotiation of the human and the nonhuman" (286). Of course, Bogost is writing philosophy, not "literature." Still, though he is well aware that his attempts to inhabit the "native logics" of a "flour granule, firearm, civil justice system, longship, fondant" (125) are merely instances of "speaking into the air" (as Peters would put it), Bogost nevertheless pursues his project as a defiance of the "tradition of human access that seeps from the rot of Kant" (*Alien Phenomenology* 4). As Bogost suggests, "metaphysics need not seek verification, whether from experience, physics, mathematics, formal logic, or even reason. The successful invasion of realist speculation ends the reigns of both transcendent insight and subjective incarceration" (5). This language recalls not only the emancipatory rhetoric of Blake's poetry but also, and in an uncanny way, the rhetoric and philosophy of André Breton. As forecast by the dizzying juxtaposition of found objects in the work of OOO and SR – the meeting of a flour granule and firearm on an operating table – the reification of "wonder" about the universe of objects is not only typically romantic but is also a cornerstone of surrealist philosophy and aesthetics.

Perhaps the best place to trace a link between surrealism and OOO is in Breton's quirky book *Mad Love*, in which he introduces the concept of "objective chance," exemplified by the shared discovery with Alberto Giacometti of evocative and erotic *objets trouvés* at a flea market in Paris. In Breton's words:

> the finding of an object serves here exactly the same purpose as the dream, in the sense that it frees the individual from paralyzing affective scruples, comforts him and makes him understand that the obstacle he might have thought insurmountable is cleared. (32)

Of special importance to Breton is that the objects found at the flea market were discovered *together*, by both he and Giacometti. The found object therefore achieves something that the dream cannot – a mutual experience that Breton describes as a sudden atmospheric condensation, "producing flashes of lightning" (33). This condensation, achieved by an object mediating between humans, is at once Platonic, romantic and surrealist, exhibiting the desire for contact and unity described above. Here, as Richard Coyne suggests, Breton also exhibits a prescient desire for the "holistic unitary patterning of chaos theory" (Coyne, *Technoromanticism* 249):

> It is as if suddenly, the deepest night of human existence were to be penetrated, natural and logical necessity coinciding, all things being rendered totally transparent, linked by a chain of glass without one link missing. If that is simply an illusion, I am ready to abandon it, but then it must be *proved* an illusion. Otherwise, if, as I believe, it may be the beginning of a contact, unimaginable dazzling, between man and the world of things. (40)

Breton's mad love shares with OOO an object-oriented metaphysics that "need not seek verification," a metaphysics that unchains him from the shackles of "subjective incarceration" (Bogost, *Alien Phenomenology* 5). It may seem counterintuitive to mount a philosophy of objects as a strategy against subjective incarceration, but there is nothing objective about the surrealist object, nothing categorical or taxonomical. As Richard Coyne suggests, "to valorize the notion of the decontextualized (or recontextualized) object may seem to contradict the subjective focus of surrealism, but the surrealists were interested in instantiations and not classification, objects placed in the 'wrong' categories" (*Technoromanticism* 191).

Recalling Lyotard's focus on the sublime in both romanticism and avant-garde art, what matters most in surrealism as in OOO is that the sublime object has the potential to disrupt traditional categories. Bogost gives a clear outline of this disruptive process when, citing Latour's animistic description of actants as "troubled souls that have not been offered a decent burial," he describes alien ontology as a "bestiary of the undead" (*Alien Phenomenology* 133):

> In the face of the undead, we exhibit terror. Troubled souls seek relief, silence, release. They operate by broken logics, ones recognizable as neither alive nor dead but striving for one or the other. We fear them because we have no idea what they might do next. (Ibid.)

As tempting as it is to diagnose OOO as a form of necrophilia (viz. Nathan Gale's "zombie ontology"), following perhaps Eric Fromm's apocalyptic view of technoculture as a world of machine-love, it is more productive perhaps to consider OOO as a flight from finitude, one that is less a philosophy than a form of creative writing, a *poeisis* that resists the closure of mechanical theories of "human access" (Bogost, *Alien Phenomenology* 4). To borrow the words of Stanley Cavell, OOO represents "the human effort to escape our humanness," which is delimited by both the finitude of our decaying bodies and the finitude of communication (86). And as Cary Wolfe suggests, acknowledgement of this finitude calls for a philosophy that "can no longer be seen as mastery, as a kind of clutching or grasping via analytical categories and concepts" (71). Perhaps OOO is such a philosophy; however, its immersion in wonder and the sublime do not suggest an acknowledgement of finitude, but a reveling, with varying ethical concern, in the infinite.

As Lyotard suggests, the ultimate fate of the ineffable and sublime object of surrealism was commodification, as is evident in a conspicuously taxonomic exhibition of "Surreal

Things" hosted by the Art Gallery of Ontario in 2009. Object-oriented ontology, which Bogost describes as an "event" of anticorrelationism, is an attempt to resist such closed taxonomies. Having suggested that the avant-garde's commodification marks the end of the sublime aesthetics, Lyotard proposes that "sublimity is no longer in art, but in speculation on art" (211). These words ring clear today if we consider OOO as a sublime philosophy, a philosophy of sublimation even, taking into account the various definitions of this term, from chemistry (a state of becoming), psychology (the diversion of sexual energy into socially acceptable manifestations), and finally, as a neologism, from artistic practice (the production of sublime objects). With this in mind, I would like to look carefully at Bogost's turn from written philosophical speculation to philosophy as *carpentry*.

Taking a literal suggestion from Harman's description of OOO as "the carpentry of things," Bogost suggests that "the job of the alien phenomenologist might have as much or more to do with experimentation and construction as it does with writing or speaking" (109). Armed with this nascent methodology, Bogost describes several of his digital media projects, including *Latour Litanizer*, *I am TIA*, and *Deconstructulator*, as "*artifacts*" of alien phenomenology. It is in this move toward an *applied philosophy*, I would suggest, that Bogost's work is the most productive, the most resistive to the reification, commodification, and Kantian classification of things. Unfortunately, the term *carpentry* seems to be a misnomer here, as Bogost's projects are clearly not the product of hammer and nail but of microprocessor and screen. As such, they fold alien phenomenology back into technoromantic, emancipatory rhetorics. Bogost's applied work is admirably cross-disciplinary, pointing the way toward applied philosophical modes that can help redefine the way research is conducted and disseminated in the humanities. However, his version of *carpentry*, which involves "constructing artifacts that illustrate the perspectives of objects," lacks Jane Bennett's commitment to political intervention.

Furthermore – and with all due respect to Bogost's characterization of academics as "insufferable pettifrogs who listen or read first to find fault and only later to seek insight" (*Alien Phenomenology* 91) – by limiting himself to "memory addresses and ROM data, or webpages and markup" (109), Bogost misses an opportunity to engage in a richer diversity of philosophical things, such as greenhouses, treadmills, dirt, canoes, radish seedlings, antique wedge-shaped coffins, cockroaches, retro arcade cabinets, and penny-farthing bicycles.

This self-indulgent litany emerges from projects conducted by the Critical Media Lab (CML) at the University of Waterloo, where the production of critical objects is also informed by an applied philosophical method. One recent project involved wiring a stationary penny-farthing bicycle with digital sensors so that a pedalist's speed and heart rate controlled the velocity and appearance of a projection flying across a walkway over the main street in Kitchener, Ontario. This project, entitled *Cycle of Dread*, was at once an embodiment of William Blake's own applied philosophy (the flying projection was that of William Blake's *Soul of the Strong, Wicked Man*) and an investigation of Mihaly Csikszentmihalyi's theory of cognitive *flow* (1990) as it is manifested in both physical exercise and immersion in a digital interface. Whereas Bogost calls his work *carpentry*, as I have written elsewhere, the CML adheres to a method called *applied media theory* (2012). I make the comparison between methods here to suggest, following Bogost, that *carpentry* might be extended to include other forms of investigation in philosophy and in other humanities disciplines (Bogost, *Alien Phenomenology* 109). *Cycle of Dread*, for example, is designed to resist the sedentary activity and narratives of disembodiment prevalent in technoculture by pointing backward to William Blake's radical aesthetics, and by pointing forward to new possibilities in interactive gaming that combine laboring bodies with digital spectacle. The ultimate goal of this project, however, is not to point backward and forward, but to point sideways.

Fig. 1. *Cycle of Dread*. Critical Media Lab., University of Waterloo.

The curious penny-farthing in *Cycle of Dread*, a bike without a chain, channels our attention toward the alien objects against which we bump and rub in what Bogost calls the "chaining together" of "tiny universes" (25). Recalling Jane Bennett's version of flat ontology, object-oriented methods in the humanities have the potential to draw "human attention sideways, away from an ontologically ranked Great Chain of Being and toward a great appreciation of the complex entanglements of humans and nonhumans" (111). In the case of *Cycle of Dread*, attention is drawn sideways not to revel in the infinitude of object being but for the sake of intervening in a technoculture possessed by a technoromantic idealism. This is a culture that relies on the disposability of objects, including those neurological objects in the human brain that are responsible for attention and that are the target of an increasingly invasive "attention economy" that vies for our "brain time." As Bernard Stiegler puts it in *Taking Care*, digital media, as currently wielded by the cultural industry, form a "network of *pharmaka* that have become extremely toxic and whose toxicity is systematically exploited by the merchants of the time of brain-time" (85). Rather than calling for a Luddite rejection of digital media in his *therapeutics of care*, Stiegler suggests that the cognitive toxicity of contemporary technoculture can be overcome

through the invention of a new way of life that takes care of and pays attention to the world by inventing techniques, technologies, and social structures of attention formation corresponding to the organological specificities of our times, and by developing an industrial system that functions *endogenously* as a system of care: *making care its "value chain" – its economy.* (48)

To invent a new economy of care in our technocultural system, philosophers would do well to intervene directly in that system's mode of production, namely digital production, offering alternative models for technological invention that draw attention sideways, to the complex entanglements of human and nonhuman things. Perhaps the best way to achieve this attentional mode is through an interlinking of digital media and conspicuous objects-to-think-with: penny-farthing bicycles, cockroaches, canoes, and copper plates. To complement Jane Bennett's ecopolitics of things, I am recommending a techno-noopolitics of things, one that acknowledges at once the infinite intermingling of things and the finitude of human being. Such acknowledgement is a necessary component of a therapeutics of care (*sorge*), which begins with a rejection of the flight from finitude. The object-oriented labor of love I have attempted to describe here points the digital humanities in a new direction that intervenes in the production of technoculture, vying with the cultural industry to gain a stake in that very limited natural resource known as human attention.

bibliography

Benjamin, Walter. "Paris, Capital of the Nineteenth Century." *The Arcades Project*. Ed. Rolf Tiedemann. Cambridge, MA: Harvard UP, 1999. 62–82. Print.

Bennett, Jane. *Vibrant Matter: A Political Ecology of Things*. Durham, NC: Duke UP, 2010. Print.

Blake, William. *The Complete Poetry and Prose of William Blake*. Ed. David V. Erdman. New York: Anchor, 1995. Print.

Bogost, Ian. *Alien Phenomenology or What it's Like to Be a Thing*. Minneapolis: U of Minnesota P, 2012. Print.

Bogost, Ian. "Latour Litanizer." Weblog entry. *Ian Bogost – Videogame Theory, Criticism, Design*. 16 Dec. 2009. Web. 11 June 2012. <http://www.bogost.com/blog/latour_litanizer.shtml>.

Breton, André. *Mad Love*. Trans. Mary Ann Caws. Lincoln: U of Nebraska P, 1987. Print.

Cavell, Stanley. *This New Yet Unapproachable America*. Albuquerque, NM: Living Batch, 1989. Print.

Coyne, Richard. "Cyberspace and Heidegger's Pragmatics." *Information Technology and People* 11.4 (1998): 338–50. Print.

Coyne, Richard. *Technoromanticism*. Cambridge, MA: MIT P, 1999. Print.

Csikszentmihalyi, Mihaly. *Flow: The Psychology of Optimal Experience*. New York: Harper, 1990. Print.

Gale, Nathan. "Zombies Ate my Ontology." Weblog entry. *An Un-Canny Ontology*. 17 Aug. 2009. Web. 11 June 2012. <http://uncannyontology.blogspot.ca/2009/08/zombies-ate-my-ontology.html>.

Haraway, Donna. *When Species Meet*. Minneapolis: U of Minnesota P, 2007. Print.

Harman, Graham. *Tool Being: Heidegger and the Metaphysics of Objects*. Chicago: Open Court, 2002. Print.

Latour, Bruno. *We Have Never Been Modern*. Cambridge, MA: Harvard UP, 1993. Print.

Lyotard, Jean-François. "The Sublime and the Avant-Garde." Trans. Lisa Liebmann, Geoffrey Bennington, and Maria Hobson. *The Inhuman*. Cambridge: Polity, 1991. 89–107. Print.

Martin, Julia. "The Speaking Garden in William Blake's *The Book of Thel*: Metaphors of Wisdom and Compassion." *JLS/TLW* 19.1 (2003): 53–81. Print.

Mitchell, W.J.T. "The Rights of Things." Foreword to Cary Wolfe, *Animal Rites: American Culture, the Discourse of Species, and Posthumanist Theory*. Chicago: U of Chicago P, 2003. ix–xiv. Print.

O'Gorman, Marcel. "Broken Tools and Misfit Toys: Adventures in Applied Media Theory."

Canadian Journal of Communication 37.1 (2012): 27–42. Print.

Perniola, Mario. *The Sex Appeal of the Inorganic: Philosophies of Desire in the Modern World.* Trans. Massimo Verdicchio. London: Continuum, 2004. Print.

Peters, John Durham. *Speaking into the Air: A History of the Idea of Communication.* Chicago: U of Chicago P, 1999. Print.

Plato. *Six Great Dialogues.* Trans. Benjamin Jowett. Minneola, NY: Dover, 2007. Print.

Sander, David. "'Habituated to the Vast': Ecocriticism, the Sense of Wonder, and the Wilderness of the Stars." *Extrapolation* 41.3 (2000): 283–97. Print.

Stiegler, Bernard. *Taking Care of Youth and the Generations.* Trans. Stephen Barker. Palo Alto, CA: Stanford UP, 2010. Print.

Wolfe, Cary. *Animal Rites: American Culture, the Discourse of Species, and Posthumanist Theory.* Chicago: U of Chicago P, 2003. Print.

Wolfe, Cary. *What is Posthumanism?* Minneapolis: U of Minnesota P, 2010. Print.

the banquet

In the opening chapter of his book *The Open*, Giorgio Agamben describes a thirteenth-century miniature depicting the messianic banquet where the members of humanity who remain are illustrated with animal heads. Agamben reads the image to suggest that on the last days of the world, human and animal natures will be transformed, in the sense that humans will become (like) animals, reconciled with their animal natures so to speak.

In the miniature, the guests are just about to eat together. For Agamben, this is an image of a coming community – the animal-headed creatures at the table represent different parts of the animal kingdom – the eagle, the ox, the lion, the ass and the leopard – bounded by the act of eating. Two musicians, also with animal heads, entertain the guests; one plays a fiddle and has the face of a monkey. In coming together to enjoy music and food, new sets of relations emerge between animals and humans, and, Agamben speculates, between animals themselves.

> [T]he idea that animal nature will also be transfigured in the messianic kingdom is implicitly in the messianic prophecy of Isaiah 11:6 [...] where we read that "the wolf shall live with the sheep, / and the leopard lie down with the kid; / the calf and the young lion shall grow up together, / and a little child shall lead them. (*The Open* 3)

The little child is a reminder that the messianic banquet is a Bible story presented to us in the form of a Picture Book. If we attend to this frame, the scene may be read as an invitation to imagine, with the child, how the guests might be rearranged through the act of feasting

stephen loo
undine sellbach

A PICTURE BOOK OF INVISIBLE WORLDS
semblances of insects and humans in jakob von uexküll's laboratory

together. In particular, one might begin to wonder about those members of the animal kingdom not included in the picture. There are no insects, for example, or other invertebrates recorded at the banquet. Yet once one reflects on this absence it begins to feel as though these small creatures might already be there – under the table, in the food perhaps and certainly in the guts of the guests. It is also imaginable that in anticipation of salvation, the stomachs of the righteous fill with butterflies, their thoughts buzz and their skin crawls. This suggests that new sets of human–animal relations are not merely cognitive ones – they are felt and imagined by different parts of the body.[1]

Agamben does not ask whether there might also be insects attending the banquet. Nevertheless, the main scenes of eating he goes on to describe in *The Open* all involve insects and other invertebrates in a laboratory setting.[2] This shift is striking. If, as Agamben imagines, the table is the place where the relations between humanity and animality could one day be reconfigured, then the laboratory seems to be the place where the distinction between human subject and animal behaviour is most powerfully drawn. In one laboratory experiment in *The Open*, a bee is placed in front of a cup of honey. As the bee begins to drink, its abdomen is cut away, but yet it is observed to keep sucking as the honey flows out of its open stomach (52). As Agamben points out, this experiment is used by Heidegger to distinguish the instinctive captivation he believes is typical of animal behaviour, from the openness particular to human existence.

Yet by placing these two scenes of eating side by side, Agamben may be doing more than drawing a contrast between the division of human from animal as enacted in the laboratory, and the new conjunctions he imagines at the messianic table. The experiments that Agamben describes in *The Open* belong to the unconventional biologist Jakob von Uexküll, who broke with the scientific paradigm of the time by refusing to study animals as a series of isolated behavioural traits. Drawing on careful observations of animals interacting with their environments, Uexküll set out to intuit the lived worlds of the small creatures he worked with. For Uexküll, genuine biological investigation entails a certain willingness, on the part of the scientist, to evoke in the "mind's eye" what is forever inaccessible to our physical senses – the radically different spatial, perceptual, temporal and affective worlds of other animals. Uexküll, as we will argue, can be understood as a pioneer of ecological thinking, because the observational and imaginative techniques he develops in the laboratory for attending to insects and other invertebrates have the potential to reconfigure traditional hierarchical divisions towards new conjunctions between animals, humans and environments.

Dorion Sagan, in his introduction to the new translation of Uexküll's *A Foray into the Worlds of Animals and Humans and A Theory of Meaning* observes: "at one and the same time Uexküll is a kind of biologist-shaman attempting to cross the Rubicon to non-human minds, and a humble naturalist closely observing and recording his fellow living beings" (Uexküll, *A Foray* 20). Here, Uexküll is understood in two contrasting ways. On the one hand, we see the "humble naturalist" pre-empting current research into animal perception and emotion, biosemiotics,[3] and the agency of self-regulating systems. On the other, the "biologist-shaman" gestures to a transcendental realm, where the complex web of relations lived by different organisms plays out in a vast "symphony" of Nature (ibid. 189).

What neither of these readings adequately recognizes is that for Uexküll there is always something fabulous, fabricated and child-like about the whole enterprise of reconstructing the subjective environments of the small animals he works with. Indeed, the subtitle to the original German *Streifzüge durch die Umwelten von Tieren und Menschen* (retained in Claire Schiller's original translation "A Stroll through the Worlds of Animals and Men," but missing in Joseph D. O'Neil's newer translation *A Foray into the Worlds of Animals and Humans*) is *Ein Bilderbuch unsichtbarer Welten* [*A Picture Book of Invisible Worlds*].[4] In so doing, Uexküll imagines the co-effecting relations between different animal environments in ways that are not fully captured by the two conventional readings. By attending to the Picture Book as a technic for ecological thought and imagination, we intend to revisit the conjunctions between humans, animals, the laboratory experiment and the festivities of music and eating, and consider whether the small creatures that Uexküll describes may in some way enable the emergence of new ethical sensibilities and relations.

the laboratory experiment

The field of animal studies has critiqued the laboratory experiment for its reductive picture of the animal. Through the tools, techniques and spaces of the laboratory the scientist is established as distant, dispassionate observer; the animal is separated from its environment for the purposes of isolating its behavioural traits, instincts or physiology, and the suffering inflicted by these experiments is ignored or deemed secondary to human-centred outcomes. Most research in animal ethics has focused on human obligations to the larger animals via shared attributes such as the capacity to suffer and feel pleasure, self-awareness, the face-to-face encounter, or the social bonds such as love or care.[5] These studies challenge the reduction of varied animal behaviours to fixed mechanical drives and affirm a complex web of differences and affinities between humans and other animals.[6] On this basis, ethical codes have been gradually devised to regulate the ways in which mammals and other larger animals are used in laboratory experiments. Insects, however, are not included. Tiny, multitudinous and almost machine-like, seemingly with limited recognizable emotion or self-consciousness, they do not register easily as objects of moral consideration or agents of ethical change. In his landmark anthology *Insect Poetics*, Eric C. Brown argues that "the insect has become a kind of *Other*, not only for human beings but for [...] animals studies as well, best left underfoot or in footnotes" (6).

What happens when we consider that insects are also present in the laboratory? In *The Open*, Agamben uses experiments on invertebrates to identify a general conceptual apparatus he calls the *anthropological machine*, which isolates biological life as an object of management and study within and outside the human (33). According to Agamben, the development of modern science was made possible, in part, because of the creation of a strategic divide between biological life − life conceived as blind instinctual acts of assimilation and excretion, and relational life − life capable of perceiving and acting out the relation between self and other, interior and exterior, incorporation and exclusion. The mechanism for this divide is a tendency to presuppose *the human* every time biological life is defined. Agamben

names this "the anthropological machine," because in the absence of any clear trait capable of defining human beings it continuously enacts the distinction, with *the animal* produced as the excluded by-product of human self-definition. In order to "oppose man to other living things," Agamben thinks that the divide must first pass within the human, isolating the animal in the human being, so to speak (15). His aim in *The Open* is to better understand how this mechanism works, in the hope of countering the damaging effects of its continuously shifting decision between life that matters and life outside all ethical consideration.

Reflecting on this, it is possible to see why insects and other invertebrates are so productive for Agamben in his analysis of the anthropological machine. The double otherness of insects in the laboratory − animals are other than human, insects are other than animals − draws attention to the *mobility* of the mechanism by which we define the human, and the fact that this is a strategic rather than a natural divide. Insects are simultaneously distant and proximate to us − we use their external behaviours to identify the base animal instincts inside us − and in this way a line is drawn inside *and* outside the human being. Building on Agamben, we can say that when qualities pertinent to human ethics, such as "self-awareness" or "sentience," are extended to other animals, the anthropological machine may not necessarily be disenabled; rather, the divide between ethical and instinctual life is redrawn in a new way. As Agamben observes, often in the case of "higher" mammals as with human beings, we imagine that two modes of existence seem to inhabit the one body − an organic life and a life that consciously negotiates its relation to an outside. The "lower" animals, on the other hand, seem to exhibit only a set of blind instinctual traits (14).

uexküll's laboratory

In this context, Uexküll's unconventional laboratory is of particular interest. He does not extend subjective concepts and feelings from a human-centred world to these small creatures, but nor does he interpret them in terms of fixed drives and external traits. Instead, he posits that every small animal has an Umwelt – a unique "foreign" subjectivity of its own. By provoking his readers to consider the lived worlds of simple instinctual organisms – the grasshopper, the tick, the hermit crab, the jellyfish, the housefly and the snail – Uexküll confounds the divide between subjective relational life and blind organic existence that is drawn by the anthropological machine. As Sagan writes in his "Introduction" to *A Foray into the Worlds of Animals and Humans*,

> the phenomenon might be described as the return of the scientifically repressed: what is excluded for the sake of experimental simplicity eventually shows itself to be relevant after all [...] With Uexküll the inner real comes back in the realization that not only do we sense and feel, but so do other sentient organisms; and that our interactions and signalling perceptions have consequences beyond the deterministic oversimplifications of a modern science that has bracketed off all causes that are not immediate and mechanical. (8)

For Uexküll, not only do different animals experience the world in different ways but sometimes there are even multiple subjectivities in the one animal. So, for example, the sea urchin has no central organization, and on this basis Uexküll speculates that skin, spines, legs and claws must each possess their own perceptual universe (*Foray* 77).

As Agamben points out, Uexküll's re-creations of the different worlds inhabited by non-human animals have a deeply disorienting effect on the world of the reader, "who is suddenly obliged to look at the most familiar places with non-human eyes."[7] But Uexküll goes a step further, insisting that in the subjective environments of other animals the whole milieu of the laboratory experiment is reconfigured. Although he uses many of the classic devices of his time – clamps, wheels, bell jars, trolleys, partitioned spaces, recording devices, diagrams, artificial membranes, grafts and dissections – he insists that no laboratory tool can ever function as a constant in animal experiments. Matter changes its

form, composition and meaning, depending on how it is perceived, used or ignored by each animal. "In this human environment, matter is the *rocher de bronze* on which the universe seems to rest, yet this very matter volatizes from one [animal] environment to another" (*Theory of Meaning* 198).

In spite of the potentially *volatizing* effects of the radically different subjective worlds of the small animals in his experiments, we also see that the conception of Umwelt is *affected back* by the tools that are used in its laboratory re-creations. As Geoffrey Winthrop-Young observes in his "Afterword" to *A Foray into the Worlds of Animals and Humans*, the role of tools in the evocation of animal Umwelt is often a blind spot for Uexküll. "Uexküll constantly denounces machines but then resorts to a Helmholtz world of cycles, coupling and feedback routines to describe the subject's Umwelt wiring" (238). Here, the observational techniques, categorizations, and metricization of the laboratory are so powerful in their tending towards a complete and measured picture of the animal that they seem to foreclose what humans can observe and think.

So, for example, reflecting on his re-creations of the life-worlds of the invertebrates he works with, Uexküll concludes that it is a general characteristic of the concept Umwelt that every animal exhibits a close functional unity with its environment ("A Stroll" 6). In his most famous re-creation, the subjective environment of the tick is reduced to three carriers of significance, which are highly selective samples taken from our more complex human environment: butyric acid – the odour of sweat common to all mammals, hairy skin covered in blood vessels and liquid at the temperature of 37 degrees.[8] The job of the researcher is to identify these markers.[9] As Agamben writes:

> Everything happens as if the external carrier of significance and its receiver in the animal's body constituted two elements in a single musical score, almost like two notes on the "keyboard on which nature performs the supratemporal and extraspatial symphony

of signification," though it is impossible to say how two such heterogeneous elements could ever have been so intimately connected. (*The Open* 41)

Again, the humble naturalist is also the biologist-shaman. As Agamben speculates, if we re-imagine the world with the *tick at its centre*, then it must be

> immediately united to these three elements in an intense and passionate relationship the like of which we might never find in the relations that bind man to his apparently much richer world. The tick is this relationship; she lives only in it and for it. (Ibid. 46–47)[10]

At once dedicated naturalist and biologist-sharman, Uexküll identifies the "task" of the ecologist to reach the "limits" of animal worlds. But what techniques does he use to attend to these limits? In the preface to "A Stroll through the Worlds of Animals and Men," Uexküll describes his method: "This little monograph does not claim to point the way to a new science. Perhaps it should be called a stroll into unfamiliar worlds; worlds strange to us but known to other creatures, manifold and varied as the animals themselves" (5). His investigations of the limits of the "dwelling-worlds" of other animals will take the form of a collection of imaginative forays, a reading that is affirmed by the original subtitle of his book: "A Picture Book of Invisible Worlds."

Now if we look closely at "A Stroll through the Worlds of Animals and Men" and Joseph D. O'Neil's more recent translation, we can see that both versions of Uexküll's essay are quite literally Picture Books, with illustrations (in black and white and colour) and chapters titles that invite the reader to experiment with shape, colour, form, space, counting, time and movement. In the drier titled *A Theory of Meaning*, the devices of the Picture Book are equally present: there is a spider that is likened to a blind tailor (158), household knives and forks are made strange through the eyes of a dog (142), and a carnivorous triton that receives a "great surprise" when sinking "its sharp teeth" into "its writhing prey," only to discover that it has grown vegetarian gums (153).

The Picture Book quality of Uexküll's writing is perhaps what helps it sustain so many contrary possibilities – the de-centring Umwelt and the closed Umwelt,[11] the "humble naturalist" and the "biologist-shaman." Uexküll invites his readers to give over to a set of self-evident instructions about animals and their relationship to the natural world, in a manner suggestive of the child's passage from a fairytale world into an unfolding science. But at the same time, like any good storyteller, he diverges from the idyll, only to enter more strangely and deeply into it. What happens, then, if we consider the technique of Uexküll's Picture Book more carefully by imaginatively engaging with it?

THE GRASSHOPPER CABARET

One fine spring day, a grasshopper, chirping to herself strolls down a...

CORRIDOR.

At the end of the corridor
she finds a...

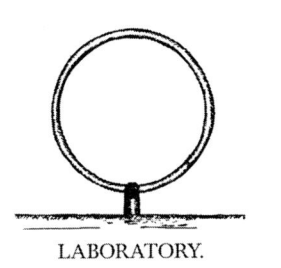

LABORATORY.

In the laboratory, she
continues to sing.

In front of her is a...

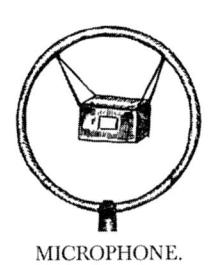

MICROPHONE.

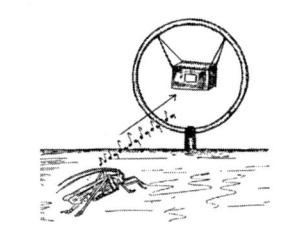

The Microphone is ON.

In a neighbouring room, two
grasshoppers, her suitors,
gather by a...

SPEAKER.

The speaker is connected to
the microphone via a long...

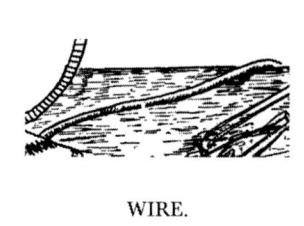

WIRE.

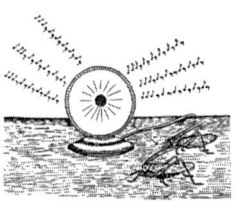

The speaker is singing to
the grasshoppers.

The grasshoppers are
spellbound, but do not chirp.

There is another grasshopper
in the room.

She is under a...

GLASS BELL.

She is also singing for her
suitors.

But alas the bell is sound
proof, and she is ignored.

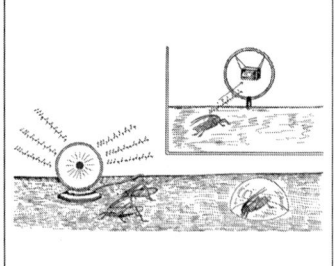

technics in the laboratory told through a picture book

"The Grasshopper Cabaret" is a story we made up as part of a series of children's writing experiments based on Uexküll's laboratory.[12] The story is a response to a picture of an experiment that Uexküll conducted on grasshoppers (illustrated by G. Kriszat) in order to demonstrate how the actions of these small animals are not goal oriented but follow the "plan of nature" inherent in their Umwelt (*Foray* 88).

Read as a conventional experiment one might say that here Uexküll is demonstrating, in a pictorial manner, the closed loop of relations between animal and environment. In order to investigate the limits of the grasshoppers' world, he uses the tools of his laboratory to break their functional cycles. By retelling this as a story in a Picture Book, we notice the presence of tools and technologies in the scene, which were there all along as techniques of seeing, but which need to remain in the background in order for the effect of the closed loop to be demonstrated.

In this context, Bernard Stiegler has provided us with a way to look upon the actual strategies and technologies of the laboratory that is not merely reductive. To Stiegler it is difficult, if not impossible, to fathom the evolution of what is the human from the evolution of "*technics*," which he defines as the exterior organized realm of inorganic matter (*Technics and Time, 1* 17). In Stieglerian terms, both the anthropological machine as a conceptual mechanism and the scientific laboratory as an actual one, upon which human self-definition rely, would be two technics amongst many in a larger technical consciousness at the "origin" of the human being.[13]

Technics allows us to see how external realms of scientific experimentation – tools, diagrams, equipment, data, languages, codes, epistemologies – are folded into the internal definition of the human. Stiegler posits that these externalities or artificial technical apparatus make possible, or in fact *are*, the retention of human experience and memory. It follows, for Stiegler, that variations in the evolution of

technical objects become *records* of transformations in human experience. Technics is the process of the "exteriorization" of human beings, whose experience is preserved in external technical objects in what Stiegler calls *tertiary memory*. Understood in Stieglerian terms as an exteriorized technics, the human being is at once the centre in human–tool relationships, but also always already de-centred, where something like a scientific laboratory becomes, as Nathan van Camp would argue, "living memory grafted onto non-living matter." Van Camp continues by saying that by focusing on co-evolution of humans and exterior organized matter such as tools, Stiegler does not address the complications posed by the presence of other animals in the laboratory setting.[14] Nevertheless, his concept of technics provides us with a way to begin to see how Uexküll's unorthodox methodology eschews the complete and measured picture of the animal that forecloses what humans can observe and think about, and with, animals.[15]

Let us now consider how technics, as a way of understanding the laboratory, can be related to the Picture Book technique that Uexküll uses to make these tools visible. The Picture Book for children is a literary genre with specific histories and audiences that differs between cultures and social contexts, which entails highly specific imaginative devices for organizing the fluid relationships between image and text, narrative and instruction, fiction and reality. If we see the Picture Book as part of an assemblage of techniques and technologies, including image making, written language and the printing press, then it becomes possible to place it alongside the scientific laboratory, and the anthropological machine which we outlined above, within a Stieglerian framework, as a technical means by which social and psychic expressions are inherited through externalizations.

According to Stiegler's argument on tertiary memory, technical externalities retain not only human experiences but also regimes of "attention." The "internalization" of attention is not something that is lived in a naturalistic sense but an intergenerational inheritance that arrives from the outside (*Taking Care of Youth* 8). In the case of the Picture Book, the modes of attention and imagination it enables require the concentration of adult attention on the juvenility of the child, something that can either be actively cultivated or discouraged.[16]

Reflecting on this, we can say that the Picture Book is a distinctive technics in that, as a mode of tertiary memory, it tends to be doubly forgotten in the world of adults. On the one hand, like other tertiary memory, it retains experiences which are inherited and not directly lived; but on the other, the experiences retained, and the kind of attention it requires, are not taken seriously by the dominant paradigms of scientific and philosophical thinking. Filled with aporias, mythologies, and hybrid objects, the childhood Picture Book is not rational in cognition but is constantly expanding with a sense of its horizons (*Taking Care of Youth* 108).

Understanding, in biology and in other fields that proceed from a techno-scientific paradigm, arrives from the construction of a knowable and "whole" object or individual. As we have seen, the laboratory retains experiences that are replete with the desire for reaching the complete object or individual. The Picture Book, however, is arguably one of the most powerful practices since the advent of writing that resists whole-object relations.[17] In fact, the Picture Book, in its requisite attentiveness, exploits the dissonance between celebrating the incompleteness or the gap between knowledge at hand and objects that are yet-to-come; and over-determining the ideal of whole object based on familiar forms and knowledge.[18]

For us, this suggests that the Picture Book can enable a thinking that *feels (for)* the type of grammar that produces partial images, rather than complete objects. This attentiveness is corporeal and affectual, for it is only through the performance and re-performance of the Picture Book that the illusions of the partial objects it creates become real and concrete instantiations. If the Picture Book is understood as a technics that tends towards incomplete imaginings, what new relations emerge between humans, animals and the tools of the laboratory?

In Claire Schiller's 1957 translation of Uexküll's "A Stroll through the Worlds of Animals and Men" the picture of the grasshopper experiment is accompanied by a lyrical description in which the scientific milieu of the laboratory becomes overwhelmed by another fantastical reading. The year is 1934, and we can imagine Uexküll's laboratory transformed into a grand inter-war cabaret club. In a closed booth, in front of a majestic radio microphone, a diva grasshopper is lost in the revelry of her own performance. In a neighbouring concert hall, well-comported suitors gather, their wings glistening multi-coloured, captivated in the acoustic world of female song emanating from a large loudspeaker. Behind them is another singer, caught in an actual glass bubble, forlorn that her cries remain unheard.

The combination of singular image with lyrical narration is a classic Picture Book technique. The use of everyday human objects and tools in an unfamiliar setting is another. And the exacerbated anthropomorphism is a third.

In this strange musical concert, sound and image have fallen out of sync. The grasshopper that sings into a microphone is heard but not seen by her suitors, the grasshopper that sings under a glass bell is unheard and invisible. The words and pictures that convey the scene to a human audience are also at odds. In the written accompaniment Uexküll provides, the two grasshopper singers are female, while the grasshoppers listening at the speaker are male. But as Joseph O'Neil, points out in his 2010 English translation, female grasshoppers do not chirp (88). It seems that in the process of telling, the story of the concert overrides scientific accuracy. O'Neil's attempt to correct this error by reversing the sex of the grasshoppers is further confounded by the fact that all the grasshoppers appear to be illustrated as anatomically male.

Here we see at once the weakness and the power of the Picture Book as a strategy for thinking about animals in the laboratory. On the one hand the narrative logic of the scene is at risk of eclipsing the scientist's commitment to empirical observation. On the other the stories that biology is capable of telling us can turn out to be even stranger than fairytales. By entering the world of the scientific experiment through the Picture Book frame we are invited to imagine the unknowable worlds of the grasshoppers through over-determined anthropomorphic frameworks that collapse logical sense to perform an alternative logic of sensation and affect. This technique de-centres the human being in the laboratory, allowing other kinds of relations to become visible. Here, we no longer see a "complete" picture of the animal Umwelt; instead, the Picture Book frame assembles, in partially realized ways, different grasshopper life-worlds, each effected in its own way by the presence of human tools and out of phase with one another. These imaginings take the form of expressions, which are speculative and incomplete, because one folds into the next.

Performed through the Picture Book, we can see that the limits of the Umwelt are not fixed but drawn and redrawn in these partial expressive ways, through the uncanny picturing of different configurations of insect, technology, human relations. Here, the tools of the laboratory are also "volatized," opening up the speculation that small laboratory animals may also in some way be de-centred by their comportment to human technologies.

picturing the technique of nature

Not only does Uexküll use the Picture Book as a technic for imagining animal Umwelts in his laboratory, he also believes that nature has picturing strategies of its own. Uexküll calls these "techniques of nature," where animals enter contrapuntally into a web of co-affectual relations with each other and with inorganic elements of their environments. Uexküll's favourite metaphor for expressing this is a vast "symphony" of nature, where the perceptual worlds of different animals interact in a way that is absolutely unknowing but perfectly in tune, like two notes harmonizing in a "musical score."[19] What is less remarked upon is that Uexküll believes that ecologists can attempt to "write the score of Nature" only because animals are already making pictures of their affectual relations with each other (*Theory of Meaning* 186). In order to consider this technique of co-picturing further, we have invented another story. It combines two flies described by Uexküll – the fly in a village street, from "A Stroll through the Worlds of Animals and Men," and the fly and the spider in *A Theory of Meaning*. Our story puts the fly and the spider, which Uexküll takes as exemplary of the technique of nature in *A Theory of Meaning*, directly into the Picture Book frame.

THE SPIDER & THE FLY

Here is a village street, seen
by the scientist Uexküll.

Here is the same village street,
photographed through a screen.

Here is the village
street seen by a fly.

Imagine a picture of a
fly, by a spider.
(You can draw it here)

A Conversation Arising from the Story of
The Spider & the Fly

The final picture still needs to be made. What to draw?

Uexküll says that spiders are good at making pictures of flies. The threads of a spider web are perfectly measured to fit the flying fly.

But the spider in the story is young and has never met a fly before. So how does the picture of the fly appear?

The spider is so affected by the missing fly, that its web becomes "fly-like." Imagine! The young spider is making a picture of something it *feels* but has *not yet lived.*

What happens next?

Imagine the fly flying down the village street.

Does it see the web?

The web is so fine the fly cannot make it out. "It is indeed a refined picture of the fly which the spider produces in its web," says Uexküll. (158)

But what does the spider's invisible picture show?

The fly of course!
It is so not (like) the fly even the fly cannot see it.

It shows the limits of the fly's world.
But it is also a picture of the fly at the limits of its world.

The fly flies into the web.
And gets caught in its own picture?

To begin with, the scientist Uexküll tries to reconstruct the fly's Umwelt with the aid of a camera, an enlarger and a screen combined with careful laboratory observations about the fly's compound eye and the markers of significance in its environment. Unfortunately, these classic scientific devices result in a picture that is "disturbing" to the human eye, so he decides to retouch the scene with "watercolours" (*A Stroll* 21). With the help of Stiegler we can say that, by explicitly making the Picture Book his technic, Uexküll allows for an intensification of tertiary memories of childhood aesthetic experience, which differ from the technics of the lab.

Entering the world of the Picture Book more fully does two things. First, a different type of comportment or attention towards nature is called for: a performance of the Picture Book, rather than distanced scientific observation of the animals. Second, inside the book, humans are placed out of the frame (even their tools are gone and they are no longer making the "pictures"). Instead, there are two others – the spider and the fly – co-picturing.

Like the tool in hand for humans, the web may be considered an exteriorization of the spider – a technique that co-evolves with the spider.[20] For Stiegler, human tools and techniques are repositories of experiences that have not been lived. In the Picture Book story the young spider remembers the fly that it has not yet met in its web. So the web is co-evolving with the fly as well as the spider. Affected by the fly, the spider intuits a picture of the fly at the limits of its world. And the fly then inherits the picture in the spider's web, an image that it does not see, but "suffers" (*Theory of Meaning* 182).

semblances of animality in the picture book

Picture this. In the Picture Book there is a spider, whose web is an extension of its body. The web is a negative picture of the fly, a counterpoint to the fly's bodily capacities, dimensions and movements. The fly cannot "see" itself in the web, but it enters into a relation with the spider's web at the limits of its Umwelt. It is with the aid of the Picture Book that we as humans can imagine the life-world of the spider through the externality of its web, and the life-world of the fly through the spider via what its web is not.

The Picture Book does not show us the complete picture of the spider or the fly but provides us with what Brian Massumi would call "semblances"[21] (as opposed to re-semblances) of the fly and spider. That is, the Picture Book makes a virtual fly and a virtual spider appear. Furthermore, the Picture Book frame swerves the scopic regime away from the thinking human eye to the respective life-worlds of the spider and the fly, to give us a feeling for a spider that "knows" from birth a fly it has not yet met; and for a fly that meets its counterpoint in the web only at its death.

On the occasion of entering the Picture Book, the spider, the web and the fly assume abstract forms, which have real effects on one another. Likewise, the grasshoppers, the chirps coming through the loudspeaker, the radio microphone and the glass bell can all be viewed as abstracted pictures of animals and technologies that nevertheless enter into actual relations.

These abstract forms are "something" not fully determinate; they are not sensed in the conventional sense, they are "non-sensed" perception, almost like a passing thought that is felt rather than understood.

In the Picture Book frame, it seems that all entities, whether human, animal or technical, reach over to another to build relations in partial but concrete ways, as if they were, to use Alfred North Whitehead's concept, "prehending" (rather than apprehending) each other. Prehension is not completely a human category; it concerns relations of casual connectedness between actual entities not determined by human teleology. It is a feeling that emerges when connecting perceptions and cognitions are transported from one actual occasion to another (18).[22] The feeling of transport imposes limits on or enables potential in the apparatus of actualization (Massumi, "Autonomy of Affect" 43). As Whitehead says, this movement of prehension from one occasion to another is *life itself*: "a living person [or any organic life] is some definite type of prehension transmitted from occasion to occasion of its existence" (18). As Uexküll's animal, human and environmental entities differentially prehend each other, we can imagine the stretching and deforming of the limits of their life-worlds, towards overlapping and involuting Umwelts.

So, the Picture Book is a Whiteheadean occasion whereby individual entities, whether human, animal or technical, partially concretize as their life-worlds enmesh. The becoming of individuals is never stable but "metastable,"[23] as resultant forms of the individual ontologically contain the tendency to reconfigure themselves, based on a constantly shifting conjunction of actual and affectual internal forces struck by external forces as organized technics. In the case of our argument, the technicity of the Picture Book, in performing the foreign subjectivities of other animals as abstractions, enables a form of thought and feeling where animality and humanity take on partially concrete shapes that are not fixed. Each time the Picture Book is performed, the individuation of entities also shifts.[24] Herein lies the great difference in comportment between the distanced scientific observer of animals, and the ecologist imaginatively engaging with Picture Books! As Uexküll writes, from the conventional viewpoint of science, "there is no mammal in itself as intuitable object, only as a notional abstraction, as a concept which we use as a means of analysis but never encounter in life. With the tick, this is completely different" (*Theory of Meaning* 179). The tick encounters the mammal as a *living abstraction*, and we the readers of the Picture Book, Uexküll goes on to imply, may also use our concepts in ways that think-feel (Massumi, *Semblance and Event* 39) the strange partial concretization of our human being that the tick's intuition entails.

The shifting lines drawn between animal and human mediated by the technicity of the Picture Book may seem to re-engage the anthropological machine as a "mobile mechanism" (as Agamben has pointed out) in ways that are not so different from the scientific laboratory. But as the process of individuation is only always partial, Umwelts fibrillate between being open and closed: open as foreign subjectivities of animals and their mutual prehendings are imagined away from human technologies; closed by the inherited tertiary memory in the technicity of the tools of the lab tending towards the whole or complete entity; and reopened through the performative dimension of the Picture Book with its infantile technics of language and image. In the anthropological machine, if it continues in some form, the drawn lines themselves are partial, perforated.[25]

The Picture Book dramatizes the grey area between the seemingly closed functional cycles of animals. It pictorially abstracts the interstices of closed functional cycles, where the edges of the Umwelt are given expression as semblances. Because we can only think-feel these grey areas through the performance of the Picture Book frame, the animal cannot completely be the excluded by-product of human self-definition because it is always already caught up in co-effectual relations of co-picturing. This abstraction of the foreign subjectivities of animals in the Picture Book laboratory is actual experience – a performance of life – a "lived abstraction."[26]

A paradox appears between Stiegler's idea of tertiary memory in human technics as inherited and therefore "not lived," and the Picture Book technic which invites the expression of tertiary retentions as "lived" abstractions in the unique forms of pictures, words, and imaginings that are phenomenologically performed.[27] By "living" the "not lived" tertiary memories through abstracting exteriorizations, which to Stiegler are ontological to the process of individuation of the human, we suggest that the Picture Book enables a radically different comportment to animal and technical others and the human-self. It increases a feeling for and between one another that reaches over to configure new ecological relations in partially concrete ways.

the picture book as a technic for ecological thinking

Uexküll's contribution to ecology is traditionally read in two contrasting ways. On the one hand, Uexküll the "humble naturalist" refuses to bracket off for the sake of experimental convenience the ways in which other animals also sense, feel and interpret the world. On the other, Uexküll the "biologist-shaman" conjures forth the radically different worlds of these animals to reveal a web of co-effectual relations between animals and their environments that he likens to the harmonies of a musical score. As we have shown, these possibilities co-exist in post-humanist readings of Uexküll, which emphasize the de-centring effects of animal Umwelten, *and* in conventional scientific readings where the animal Umwelt is a closed functional loop.

What might the Picture Book do to an ecology that oscillates between the careful observations of a naturalist and a vast "symphony of nature"? At the end of *A Theory of Meaning*, Uexküll reflects on the analogy he has drawn between the co-effecting relations that the biologist observes between organisms and environments, and the "natural score" played by the instruments of an orchestra.

If we take a glance at an orchestra, we see in each individual rostrum in musical notation the voice leading, for the instrument to which it belongs, while the whole score is on the conductor's rostrum. But we also see the instruments themselves and wonder if these are possibly adapted to each other not just in their respective tonalities, but in their entire structure, i.e., if they form a unit not just musically but also technically. Since most instruments in the orchestra are capable of producing music by themselves, this question cannot be answered in the affirmative as simply as that.

Whoever has listened to the production of musical clowns, who work with instruments that otherwise serve for making noise, such as hair combs, cow bells, and other such things, will have been convinced that one can very well play a cacophony, but not a symphony, with such an orchestra. Upon closer examination, the instruments

of a real orchestra demonstrate a contrapuntal behaviour already in their structure. (189)

It seems that we have made the symphony Uexküll hears in his laboratory into a "cacophony," an assemblage of makeshift instruments played by a Picture Book clown. But we must not forget that the grasshoppers are also "making music." For Uexküll, the grasshopper's chirps are techniques of nature, exteriorizations of its capacity to form contrapuntal relations between its organs of perception and its environment.[28] To conduct this experiment, Uexküll uses recording and amplification technology and this causes the grasshopper's song to be at once enhanced and displaced via the loudspeaker. If we imagine the concert that is performed by the grasshopper in the laboratory, then there may no longer be a preordained melody but a discordant harmony interrupted by a strange silence. By "living" through the Picture Book's performance, the "not lived" tertiary memories retained in the laboratory in its abstractions, we may speculate a memorization and further extension of the Umwelt of small animals through technical means.[29]

epilogue

One way of imagining the coming community that Agamben anticipates in *The Open* would be that in the end human beings learn to accept their (base) animality, and animals come to be accepted for their human-like qualities (their intelligence, emotions, self-awareness). But if through the Picture Book we attend to the invertebrates that are with us in the laboratory and at the table, then it no longer seems to be a matter of transposing existing qualities between human and animal but of creating an aesthetic medium in which new and unforeseen sensibilities might emerge.

THE STARLING & THE FLY

The following story is told by the scientist Uexküll.

A researcher who is a friend […] raised a young starling in a room, and the bird had no opportunity ever to see a fly, much less to catch one. Then he observed that the bird suddenly started after an unseen object, snapped it up in midair, brought it back to its perch and began to hack away at it with its beak, as all the starlings do with the flies they catch, and then swallowed the unseen thing.
(*Foray* 120–21)

For Uexküll it is obvious – the starling is so overcome by a feeding mood that it magically conjures up a fly, even though there is no fly in the room.

How remarkable, a room without flies!

And more too, the starling eats an invisible fly! Magical *and* invisible, the fly must be in a Picture Book.

Of course, the Picture Book belongs to Uexküll, but wasn't it the starling that first made up the fly?

Imagine the Picture Book fly.

What becomes of the invisible fly as it is being swallowed? What picture might it make?

notes

The figures reproduced in this paper are from Jakob von Uexküll, *A Foray into the Worlds of Animals and* *Humans* with *A Theory of Meaning*, trans. Joseph D. O'Neil (Minneapolis: U of Minnesota P, 2010); copyright 2010 by the Regents of the University of Minnesota. Originally published in *Streifzüge durch* die Umwelten von Tieren und Menschen; copyright 1934 Verlag von Julius Springer.

1 Agamben observes that the animal figures representing humanity in the miniature correspond allegorically to different parts of the human body: bones, nerves, veins, flesh and skin (*The Open* 2).

2 Both the laboratory and the table are key contested sites for animal studies. As Peter Singer and others have argued, when we bring other animals to our table as food we disregard the terrible suffering incurred through factory farming and its environmental impacts. In this context, it is a provocation to consider whether, by bringing insects to the table, we are in fact including them in some way in our ethical consideration.

3 While we are cognizant that the animal worlds of Jakob von Uexküll are thoroughly opened up by the field of contemporary biosemiotics, in this paper we are concerned with the technical ontology rather than a semiotic one of scientific experimentation when it meets the children's Picture Book, and how this may provide a re-reading of Uexküll's animal perceptual worlds, and the relationship of humans with these worlds. We would like to thank our anonymous reviewer, who reminds us that the kind of aesthetic "knowing" which the Picture Book exhibits can be read through semiotics as a recognition on Uexküll's part of the importance of iconic and indexical signs for both animals and humans.

4 The figure of the child and the picture storybook are also devices that Agamben uses to imagine new relations opening between human and animal that do not serve the project of human self-definition. But what Agamben does not consider is that Uexküll's laboratory also draws on the imaginative realm of childhood, through its Picture Book frame.

5 For example, Peter Singer's *Animal Liberation* and more recently Marc Bekoff's *The Emotional Lives of Animals*.

6 For example, Cary Wolfe's *Zoontologies*.

7 *The Open* 45. Here, Uexküll aligns with the artistic avant-garde of his day, and with recent posthumanist work on non-human agencies. But as Geoffrey Winthrop-Young warns in his "Afterword" to *A Foray* and *Theory of Meaning*:

The always present danger, of course, is that this may entail a reification of other Umwelten. The question *How can we in our world see how animals see their world?* may easily turn into the more self-interested inquiry *How can we see how animals see their world in such a way that it will change and enrich the way in which we see ours?* (235)

8 Biosemioticians would call these "markers of significance" signs.

9 It is important to remember, however, that Uexküll understands these markers to be indications of what matters to the animal, rather than the human being. It is precisely because the animal decodes its environment according to a series of markers (or signs in biosemiotics) that he likens the animal to a "machine operator" but not a "machine," which reacts to stimuli from the outside world without selectively interpreting them ("A Stroll" 7–9).

10 Agamben argues in *The Open* that Uexküll may end up re-articulating the anthropological machine in a new way that lends itself to two disturbing alternatives. Either the hierarchical division between human and animal is re-established – Heidegger, for example, uses Uexküll too in order to contrast the instinctive captivation he believes is proper to all animal behaviour, with a human openness to the world – or the divide between human and animal collapses entirely, in a way that aligns with Friedrich Ratzel's politically frightening notion of *Lebensraum*, whereby "all people are intimately linked to their vital space as their essential dimension" (42).

11 If we attend more carefully to the Picture Book frame then both the previous readings of Uexküll we have outlined are affected: Uexküll the posthumanist uses a very human, child-like technique to evoke non-human worlds. But for Uexküll the ecologist who researches the limits of the animal, his task is imagined or performed in ways that suggest that its precise outcome is open.

12 "The Grasshopper Cabaret" was performed as a children's book reading as part of the symposium presentation "Ecological Thinking through the Picture Book of Jakob von Uexküll's Laboratory" in Rethinking Behaviour and Conservation: The History, and Philosophy and Future of Ethology II, Centre for Social Inclusion, Macquarie University, Sydney, 26–28 November 2011; and as a short performance at the Architecture-Writing: Experimental

Approaches symposium, Critical Studies in Architecture, KTH Stockholm, Stockholm, 24–25 May 2012.

13 Stiegler, in *Technics and Time, 1*, in an effort to postulate the origins of the human, explains the originary "default" in the constitution of the human through the myth of Epimetheus who "forgot" to confer upon humans any special gifts, and that the current power of the human being and its ability to know, think and exteriorize is the "fire" stolen from the gods by Prometheus.

14 Although, as Nathan van Camp has pointed out, by recognizing the tertiary memories of human life deposited in externalities, the human is de-centred in ways that may disrupt the anthropological machine, we have also argued that tertiary memories deposited in the tools of the laboratory tend back towards a picture of the centrality of the human which covers over this destabilizing effect.

15 Of all the animals, insects appear to have the highest technicity, so one way to adapt the conventional reading of Stiegler to human animal relations is to say that these small creatures also operate, like the tools of the laboratory, as tertiary memories of the human being. Along these lines, Jussi Parikka's recent book *Insect Media* investigates how insect modes of organization – swarms, webs and distributed agencies – provide new ways of understanding media technology and its relationship to biology, which do not rest on the notion of individual agents or a deterministic account of technology. By reading insects via media formations, his book opens up a new and rich account of the ways in which digital culture helps to form and de-centre human beings. This is a productive approach to pursue. However, we would like to avoid a reading that reduces insects to externalizations of the human, because this potentially misses the otherness of insect life that Uexküll emphasizes, and the difficult question of their place in our ethical thought and consideration.

16 For Stiegler, the capturing and formation of attention in the young by technical and media industries threatens processes of what he calls human "maturity," and global social and cultural development (see Stiegler, *Taking Care of Youth* 7–8). By contrast, we are suggesting that the Picture Book is part of the formation of nascent attention as imagination. For Stiegler, the partial or "transitional objects" opened through childhood play are the first forms of tertiary retention that can only appear in

"transitional spaces"; spaces that "form the basis of all systems of care and nurturance: a transitional space is first and foremost a system of caring" (ibid. 15). So we might also say here that the Picture Book augurs human attention that proceeds from a performativity that involves transitional objects and spaces whereby humans are struck by the infancy of inherited technics.

17 It is important to note that the Picture Book relies on the concerted use of the mode of "picturing" that significantly pre-dates writing.

18 Building on this, we can see that what is expressed in the Picture Book, in its content, structure and framing, is organized by a unique grammatization related to infancy. Here, infancy is not a nascent version of the rules of language that "matures" into adulthood, or a pre-linguistic ineffable state. Rather, according to Agamben, "infancy" marks the threshold between wordlessness and speech: "where language stops is not where the unsayable occurs, but the other where the matter of words begins" (*Idea of Prose* 27). At this threshold, which we are arguing can be found in children's Picture Books, humans are *struck* and overwhelmed by language. This affect stems from the fact that the Real of language – its grammar, words, and image-objects – are there at their limits of sense and existence.

19 *Theory of Meaning* 188–89. Deleuze, Parikka and Agamben have developed different readings of Uexküll's musical metaphors.

20 Making a similar point, in more general terms about animal Umwelt, Elizabeth Grosz has argued that the bubble world is the projection of an animal's bodily capacities (183).

21 Massumi defines semblance as "the experience of a *virtual* reality," the manner in which the virtual actually appears (*Semblance and Event* 15–16).

22 As Whitehead says, prehensions "define the real individual facts of relatedness, a kind of transportable perception or cognition extracted from other actual occasions, or prehensions are transported by perception or cognition" (18).

23 This "process of 'individuation'" as metastable comes from Gilbert Simondon (300).

24 As Grosz says, the limit space of the Umwelt is always in construction: "Space is built up, sense by

sense, perceptual organs upon organs, forming the soap bubble, its limits, its contents" (180).

25 We appropriate here a description of the line drawn by the anthropological machine which is also metonymic of the stretched boundaries of Uexküll's bubble Umwelts:

> What I have really drawn there is an oval line, for this white chalk mark is not a line, it is a plane figure, in Euclid's-sense a surface, and the only line that is there is the line which forms the limit between the black surface and the white surface. This discontinuity can only be produced upon that blackboard by the reaction between two continuous surfaces into which it is separated, the white surface and the black surface. (Massumi, *Semblance and Event* 89)

26 Massumi, *Semblance and Event* 15. Massumi derives "lived abstraction" from a Deleuzian concept, where abstraction is nothing more or less than the performative/lived dimension of that abstraction. The act of abstraction configures the potential in that abstraction, and that potential is relayed from one abstraction to another.

27 The theoretical argumentation on the implications of this paradox for Stiegler's contention with Simondon that the process of individuation is ontologically technical as the process of psychic individuation is already a collective one (Stiegler, "Theatre of Individuation"), warrants another paper.

28 The chirping sounds – for Deleuze, the exterior territory – of the animal are related to its morphology; for example, the evolution of the development of sound-making techniques is related to the distance of mates.

29 To perform the foreign subjectivity of insects and other small animals through the Picture Book frame opens up the radical possibility of insect exteriorizations – not just insects as human exteriorizations but insects with their own tools, insects reconfiguring human tools and insects displaced by tools and techniques (human and animal) that surround them.

bibliography

Agamben, Giorgio. *Idea of Prose*. Trans. Michael Sullivan and Sam Whitsitt. New York: State U of New York P, 1995. Print.

Agamben, Giorgio. *The Open: Man and Animal*. Trans. Kevin Attell. Stanford: Stanford UP, 2004. Print.

Bekoff, Marc. *The Emotional Lives of Animals*. Novato, CA: New World Library, 2007. Print.

Brown, Eric C. *Insect Poetics*. Minnesota: U of Minnesota P, 2006. Print.

Buchanan, Brett. *Onto-Ethologies: The Animal Environments of Uexküll, Heidegger, Merleau-Ponty and Deleuze*. New York: State U of New York P, 2008. Print.

Grosz, Elizabeth. *Becoming Undone: Darwinian Reflections on Life, Politics and Art*. Durham, NC and London: Duke UP, 2011. Print.

Massumi, Brian. "The Autonomy of Affect." *Parables for the Virtual: Movement, Affect, Sensation*. Durham, NC and London: Duke UP, 2002. 23–45. Print.

Massumi, Brian. *Semblance and Event: Activist Philosophy and the Occurrent Arts*. Cambridge, MA: MIT P, 2011. Print.

Parikka, Jussi. *Insect Media: An Archaeology of Animals and Technology*. Minneapolis: U of Minnesota P, 2010. Print.

Simondon, Gilbert. "The Genesis of the Individual." *Incorporations*, Zone 6. Ed. Jonathan Crary and Sanford Kwinter. New York: Zone, 1992. 297–319. Print.

Singer, Peter. *Animal Liberation*. London: Pimlico, 1995. Print.

Stiegler, Bernard. *Taking Care of Youth and the Generation*. Trans. Stephen Barker. Stanford: Stanford UP, 2010. Print.

Stiegler, Bernard. *Technics and Time, 1: The Fault of Epimetheus*. Trans. Richard Beardsworth and George Collins. Stanford: Stanford UP, 1998. Print.

Stiegler, Bernard. "The Theater of Individuation: Phase-Shift and Resolution in Simondon and Heidegger." Trans. Kristina Lebedeva. *Parrhesia* 7 (2009): 46–57. Print.

Uexküll, Jakob von. *A Foray into the Worlds of Animals and Humans* and *A Theory of Meaning*. Trans. Joseph D. O'Neil. Minnesota: U of Minnesota P, 2010. Print.

Uexküll, Jakob von. "A Stroll through the Worlds of Animals and Men: A Picture Book of Invisible

Worlds." *Instinctive Behavior: The Development of a Modern Concept.* Trans. Claire H. Schiller. New York: International Universities P, 1957. 5–80. Print.

Uexküll, Jakob von, and G. Kriszat. *Streifzüge durch die Umwelten von Tieren und Menschen: Ein Bilderbuch unsichtbarer Welten.* Sammlung: Verständliche Wissenschaft, Berlin, 1934.

Van Camp, Nathan. "Animality, Humanity and Technicity." *Transformations: Journal of Media and Culture* 17 (2009). Web. <http://www.transformationsjournal.org/journal/issue_17/article_06.shtml>.

Whitehead, Alfred North. *Process and Reality: An Essay in Cosmology.* 1929. New York: Free, 1978. Print.

Wolfe, Cary, ed. *Zoontologies: The Question of the Animal.* Minneapolis: U of Minnesota P, 2003. Print.

One thing about which fish know exactly nothing is water, since they have no anti-environment which would enable them to perceive the element they live in.

McLuhan and Fiore, War and Peace in the Global Village *175*

I dog days

In his 1970 collaboration with Wilfred Watson, *From Cliché to Archetype*, Marshall McLuhan characterizes the domestic dog as a "primordial" technology which functioned as an early extension of human sensory life, with important survival benefits for human society.[1] Like the printing press and automobile that came later, the authors argue, *Canis lupus familiaris* made possible for *Homo sapiens* a new mode of living.[2] McLuhan and Watson suggest that earlier technologies remain evident in modern times in the "verbal residues" of well-known phrases and idioms: "His bark is worse than his bite," "Every dog has its day," "A bone of contention," et al.[3] Any technology will imprint such clichés on the language, expressions that originally derived from unique and innovative practices but which soon became hackneyed and stereotyped through their endless repetition. Today they are employed through habit and convention, their origins largely unnoticed. McLuhan and Watson are keen not to confine this notion of the cliché just to language, however, which, they argue, is but one technology amongst many. All technologies serve initially to enlarge a culture's scope of action and patterns of association and awareness, only to produce environments which subsequently "numb our

tom tyler

NEW TRICKS

powers of attention by sheer pervasiveness."[4] Like the pencil or the telephone, the most revolutionary innovations soon become ubiquitous and commonplace. The social and psychological environments which result, comprising the technological and perceptual media within which we exist, are overlooked, unacknowledged and effectively invisible. "[O]ur perceptions themselves are clichés patterned by the many hidden environmental structures of culture."[5]

Throughout his work, McLuhan sought to focus on the far-reaching effects of technologies, and to highlight the invisible environments they created. In his writings and presentations he attempted always, by means of provocative prompts, probes and jokes, to draw to our attention the perceptual and conceptual clichés we so often fail to notice. In this essay I would like to explore the ways in which an unassuming and seemingly frivolous digital game requires its players to confront a longstanding and deep-seated cliché, a distinct strain of anthroponormativity that quietly but persistently discourages perceptual identification with other forms of animal life. I wish to look too at a particular, innovative way in which this game engages its participants in the process of play, evoking modes of affinity and involvement that are unbound by questions of species identity. On this unlikely platform the technological and the animal meet, and play together, to educative, enlightening effect. It is often said – it is a cliché, in fact – that you can't teach an old dog new tricks, and so I would like to suggest that we can best understand the game, and learn from its creative coupling of techne and animality, by boning up instead on a pair of established but largely forgotten conceptual ploys which remain, nonetheless, profitably challenging. In revisiting, and, perhaps, teaching to a new audience these old tricks, it is my intention to pursue thereby the essay's twin themes of alterity and identity. Like many others, our digital game depicts varied modes of interaction between members of different species, but it also extends to its players an invitation both to appreciate a certain, striking form of animal otherness and to engage, at the same time, in interspecific identification.

II a shaggy dog story

Dog's Life for the PlayStation 2, created by Frontier Developments and published by Sony Computer Entertainment Europe in 2003, is a digital game aimed principally at a children's market.[6] Players control Jake, a young, care-free, mixed-breed mutt, who, clearly without owner or primary care-giver, gets along by means of his winning ways, and by scavenging from trashcans and pilfering unattended food. At the beginning of the game, roguish Jake immediately finds himself called to adventure: "Hi, I'm Jake. Welcome to my world! Normally I'd love to stop and play ball with you, teach you a few tricks maybe, but my friend Daisy has just been dog-napped!"[7] And so it begins. Players are able to enter Jake's world and to engage in all manner of stereotypically canine behaviour, designed to appeal to the younger gamer: digging up bones, marking territory, farting, defecating, swimming and then, inevitably, shaking dry. As Jake himself says, "No school, no chores, no clothes! It's great being a dog!" Canine life is not just a matter of tug-of-war and chasing chickens, however. Players must keep Jake healthy and well fed, lest he become tired and listless, whilst avoiding at all times the dog catcher and his Doberman. And, of course, there's that matter of rescuing Daisy. Jake must track her down, searching first through the quiet corners of rural Clarksville, then on to Lake Minniwahwah ski resort, and arriving, finally, at the perilous streets of Boom City. Ultimately, he must face the elusive Miss Peaches, purveyor of the suspiciously sourced Crunchy Cat Food.[8]

The narrative of *Dog's Life*, though engaging enough, is far less interesting than a central, novel element of its gameplay. When the game begins, it is a bright, sunny day, and Jake finds himself at the edge of a flowering meadow, buzzing with insects and fluttering with butterflies (Fig. 1). As he enters, bounding after a young human companion, players are invited to switch to "Smellovision" and truly step into Jake's world. Smellovision has two key features. Firstly, play shifts from a third-person perspective which follows Jake from some distance, to a subjective, first-person perspective that permits gamers to "see the world through Jake's eyes."[9] The environment is depicted as if through a wide-angle, fisheye lens, and colours become noticeably muted, simulating a canine point of view. Secondly, players are now able to "see smells," which are represented as brightly coloured clouds of scent, free-floating or billowing around the humans, parrots, pigeons, fox, and other animals whom Jake encounters, and who leave odorous wisps in their wake. Footprints glow and can be tracked, and the locations of buried bones are revealed by shafts of celestial

Fig. 1. Third-person perspective. Players see Jake as he interacts with his environment.

Fig. 2. Smellovision. Players can "see the world through Jake's eyes," including clouds of scent, glowing footprints, and the location of buried bones.

light (Fig. 2). The game becomes, in effect, an interactive manifestation of Jakob von Uexküll's much-discussed monograph *A Foray into the Worlds of Animals and Humans*, as players are invited to step into the digital soap bubble that represents Jake's *Umwelt* (self-world). Just as Uexküll describes, the familiar meadow is transformed, and a new world comes into being.[10] I would like, initially, to focus on this playful simulation, and to examine the variety of ways in which Smellovision seeks to reproduce a canine phenomenal world.[11]

III how does your dog smell?

Smellovision effectively captures something of the greater *field of view* of canine vision. The visual field varies from breed to breed, and indeed from individual to individual, depending on the placement of the eyes in the skull. Typically, a dog's eyes deviate approximately 20° lateral to the midline, with each eye providing a monocular field of view between 135° and 150°, combining to create a total field of view of approximately 240° or 250°, allowing for overlap. A human's eyes, by contrast, both oriented straight ahead, provide a smaller field of view of approximately 180°. The fisheye lens of *Dog's Life* is thus a means of communicating Jake's greater visual field within the constraints of a conventional television screen. Similarly, the significantly lower *visual perspective* that dogs normally experience is replicated by the game. In shifting to Smellovision, players find themselves just inches from the virtual ground, confronted in the first instance by car wheels, trash cans and the legs of humans: you must actively look up in order to see the faces that now tower above you. Further, although it was believed for a long time that dogs lacked all *colour vision*, this is not, in fact, the case. The retina of a human eye ordinarily has three kinds of photoreceptor cells that detect colour, called cones, each of which is especially sensitive to a different wavelength of light: red, green and blue. In combination, these three types of cone make possible trichromatic colour vision, the detection of a vast range of colours. A dog's eye, on the other hand, contains just two kinds of cone, which respond most to violet and yellow-green wavelengths. As dichromats, dogs thus detect a much smaller spectrum of colours than most humans, and cannot distinguish, for instance, red from green, or blue from violet. This colour blindness is carefully simulated by the subdued palette of Smellovision, which provides the player with a reduced range of colours, in dull shades.[12]

As a full simulation of the alterity of canine vision, however, *Dog's Life* ultimately falls short. The game does not attempt to demonstrate dogs' limited *depth perception*, for instance. Precisely because their eyes are set further to the sides, affording a wider field of view, dogs have a smaller area of overlap in the middle, and thus less three-dimensional vision, which amounts to no more than 60°, and often much less, compared to a full 140° in humans. Dogs' visual *acuity* is similarly poor. In a canine eye, multiple photoreceptors are attached to a single ganglion, the nerve cell that transmits signals from eye to brain, increasing overall sensitivity to light but decreasing the degree of detail that can be conveyed. Canine visual acuity has been estimated to be equivalent to 20/75 human vision, which is to say that Jake would need to be just twenty feet away in order to distinguish the detail of an object that one of his human companions could see at seventy-five feet. Perhaps unsurprisingly, this aspect of canine vision is not explored in *Dog's Life*. Nor, indeed, is dogs' relatively poor *accommodation*, the ability to focus on objects situated at different distances, which renders as a blur anything closer than thirty centimetres or so from their faces. A human adult, by contrast, can on average accommodate objects as close as ten centimetres away, and a young child can usually see clearly at just five centimetres away. (This is most likely why dogs often seem not to notice objects that are right in front of them.) Finally, nothing is made during the course of the game of dogs' impressive *crepuscular vision*. A range of physiological factors provide dogs with sight which is ideally suited to the dim light to be found at dusk and dawn. Whilst the central 25° of a human retina is devoted principally to cones, providing that glorious trichromatic colour vision, in dogs this area is largely given over to rods, photoreceptors sensitive to low levels of light. In addition, a canine eye contains a layer of cells at the back of the retina, called the tapetum lucidum, which reflects back light that has already passed through the retina, effectively providing the photoreceptors with a second opportunity to capture each photon of light. Further, tapetal riboflavin may even absorb light in the shorter blue wavelengths, shifting it to a longer wavelength more easily detected by rods, and thereby enhancing the contrast between dark objects in the environment and the brighter night sky behind.[13] The ability to see in the dark, when swapping to Smellovision, might perhaps have opened up interesting gameplay opportunities, but Jake's adventures take place entirely during daylight hours.[14]

Dogs, famously, have highly sensitive noses, and *Dog's Life* works hard to explore the experience of living in a richly odoriferous world. In the first instance, with the change over to Smellovision, aspects of the environment that are entirely *invisible* to humans become apparent. Though the hues of the visible world are now muted and dull, representations of scents and smells are brightly, vividly coloured: objects as diverse as misplaced axes and wind-snatched sheet music emit swirls of vibrant scent; varied puffs of inviting odour float around every part of the environment, to be "collected" by the player; and humans and other creatures are surrounded always by a warm cloud of their own scent, a conspicuous current of oils and dead skin cells, evaporating sweat and bacteria. Dogs' ability to *locate hidden objects* is similarly integrated into the gameplay, with periodic digging necessary to unearth coveted bones, whose location is disclosed by spectacular, unmissable pillars of light. *Urine marking*, the placement of canine calling cards, is ordinarily undertaken to indicate a dog's identity, interest in mating, etc. Jake can be made to pee at any time, but in the special "pee marking" mini-game, initiated after collecting yellow scents, you must capture areas by peeing over your opponent's marks, in a contest for territory. The capacity to *track* other creatures, long employed to human advantage, depends on a dog's being able to detect when individual footprints were laid down: the odour of any given print will be fractionally stronger than its immediate predecessor, thereby revealing direction of travel. In *Dog's*

Life, trails of bright yellow shoe prints, including those left by a masked robber you must hunt down, clearly indicate direction by their distinct heel-and-sole shapes. Finally, *Dog's Life* captures not just the passive reception but also the *active pursuit* of scents. Dogs' enthusiastic, noisy sniffing, significantly more effective than the paltry human equivalent, is a complex process that simultaneously draws new air into the nose whilst expelling the old, and which in turn causes a slight wind current that facilitates further inhalation. Jake can't help but stumble into a mass of floating odours when in Smellovision, but, by pressing the appropriate controller button, players can also have him sniff at will, which causes him to turn towards the nearest scent.[15]

Effective as these many simulations of canine olfaction are, however, they do not begin to capture the astonishing extent to which a dog is sensitive to the slightest of smells, and capable of discriminating between near identical odours. Each nostril of a dog's nose draws in a distinct sample of odour, providing differential information, and research indicates that the canine sense of smell works with the exhalation as well as with the inhalation of air. The inside of a human nose contains approximately six million receptor sites, each with hairs capable of catching chemicals of a particular molecular shape. A canine nose, by contrast, contains two or three hundred million such receptor sites, and sensitivity increases exponentially. Dogs, as a result, are receptive to *minute quantities* of odour, the equivalent, for instance, of being able to detect a single teaspoon of sugar diluted in a million gallons of water. In one series of tests, dogs were able to identify a glass slide by the scent from a solitary human fingerprint with which it had been marked three weeks earlier. Other research indicates that dogs can even smell the chemicals produced by cancerous cells: from tissue or urine samples, or the patients' exhalations, dogs have accurately detected skin, breast, bladder and lung cancers, in one case when melanoma was present in only a fraction of cells and had been missed by pathological examination. Although Smellovision makes present odours that are entirely imperceptible to humans, represented as distinct, unambiguous clouds, it does not explore the extent to which dogs can detect the faintest traces of scent. Similarly, the game does not address dogs' celebrated powers of *discrimination*. Presented with T-shirts that had been worn by identical twins, dogs were able to ascertain the garments' owners, provided the shirts were placed in close proximity during the test, thereby allowing them to be sniffed simultaneously. Although different species of creature within *Dog's Life* give off distinctive odours – synaesthetic consistency within Smellovision ensures that humans are always purple, other dogs are orange or blue, parrots are green, etc. – individuals of the same species or breed are identical in terms of their scent.[16]

Dog's Life, then, is an engaging and very effective simulation of canine perception, but it is certainly not comprehensive.[17] Gameplay necessarily comes first, and the demands of the medium, genre and market prevent a more detailed or exhaustive re-creation of contemporary research into dogs' visual and olfactory capacities. Shortcomings in terms of verisimilitude, however, by no means detract from *Dog's Life*'s unrivalled capacity to highlight and bring to the fore the alterity of this particular, non-human mode of awareness and apprehension. Despite its inevitable limitations, Smellovision embodies, as we will see, the digital game's anti-environmental potential.

IV dogged determination

In welcoming us to his world, Jake expressed the desire to teach us some tricks. Caught up in his rescue mission to save Daisy, this activity seems to be postponed indefinitely, but in fact, over the course of the game, Jake is a very effective instructor. In this section and the next, I would like to explore two such tricks, two conceptual ploys, neither entirely new but both largely forgotten today, which Jake ably demonstrates if we take up his challenge and play *Dog's Life*. The first of these is one of McLuhan's lesser-known probes. As we saw, McLuhan argues that, like verbal clichés,

although new technologies may at first be startlingly innovative and invigorating, they soon become, through repetition and ubiquity, numbingly environmental. These encompassing, involving environments, he argues, are not passive wrappings but active processes that structure our actions and awareness without our noticing them. The ground rules, the configurations, the pervasive patterns of these active environments elude easy perception; they are, for the most part, invisible.[18] We are oblivious to our surroundings, immersed, like fish in water, in a medium whose significance remains unknown and unacknowledged. In order to discern the environments in which we habitually exist, to become aware of the enveloping "climates of thought and feeling,"[19] we need to be exposed, McLuhan suggests, to what he calls *anti-environments*.

Anti-environments, according to McLuhan, can promote awareness and pattern recognition.[20] They can provide new strategies of attention that train perception onto the unnoticed environment. McLuhan was fond of Hans Christian Andersen's tale of the emperor's new clothes as a means of illustrating this process by which environmental norms are disrupted and thereby brought to our attention. The "well-adjusted" ministers and officials, immersed in court culture, saw the emperor as beautifully appointed, he suggests, whilst it took an "antisocial" brat, unaccustomed to this environment, to alert everyone to what was going on.[21] It is, McLuhan argues, most often the ill-adjusted individual, the outsider, even the criminal or enemy of society, who is best placed to draw attention to the environment. Beyond the delinquent child,[22] it is to the artist, for instance, that McLuhan most often turns for anti-environmental observations. Pop Art, he argues, takes banal objects from our daily lives and reminds us that we are surrounded by a world of images and artefacts that are intended not to train perception and awareness but to produce effects for the economy.[23] Such art can elicit a "sting of perception" or "shock of recognition."[24] Similarly, McLuhan suggests, whilst the professional will tend to classify and specialize, adopting the ground rules provided by the mass response of his colleagues, the amateur, who works alone and can afford to lose, seeks instead a total, critical, anti-environmental awareness of the individual and of society.[25] The sleuth or gumshoe of popular fiction, such as C. Auguste Dupin and Sherlock Holmes, Mike Hammer and Philip Marlowe, turns a self-conscious attentiveness on the big city, "detecting the social environment by probing and transgression."[26] And, like the wisecracks and witticisms of Chandler's private investigator, McLuhan argues that jokes more generally can be our most appealing anti-environmental tool, as the funny man, an outsider with a grievance, probes into the cultural matrix that plagues him, providing as he does so a guide to changing perceptions.[27]

In addition to these inquisitive, unconventional individuals, amongst whom McLuhan himself might well be included, all manner of processes and products, many of them more mainstream, can function anti-environmentally. McLuhan recounts how Ovid's multiple plots and subplots, or succinct Japanese haiku, or successive literary movements from realism to the romantics to the modernists, can produce anti-environmental effects.[28] Schools, in turn, have traditionally been designed as anti-environments, "to develop perception and judgement of the printed word,"[29] though they are increasingly ineffectual, McLuhan argues, in today's electronically mediated milieu, which turns the whole world into a "classroom without walls."[30] Liberal studies have long been considered a means of providing orientation and perception, but when the arts and sciences themselves become environmental, new controls must be found.[31] Diverse events and artefacts – from participatory happenings[32] to the hard drugs of Burroughs' anti-Utopian novels,[33] from the jolt of bad news[34] to the very Earth itself when experienced from a post-sputnik perspective[35] – can produce anti-environmental effects in the right environmental context.[36] Even a new technology can itself enjoy a "brief reign" as an anti-environment, highlighting through sheer novelty and innovation some features of the old environment, before it becomes, inevitably, environmental itself.[37] Anti-environmental means of perception "must constantly be renewed in order to be efficacious,"[38] as the soup cans that are appropriated for the gallery swiftly become postcards and tea-towels, before being redeployed, newly formatted, to sell soup once more.[39] Environment and anti-environment alternate their roles "with all the dash and vigour of Tweedledum and Tweedledee,"[40] performing an endless, cyclic, "technological fugue."[41] Even the most popular arts, McLuhan suggests, can serve to increase awareness, at least until they become entirely environmental.[42] Games, in fact, by transforming the customary, working environment into model, paradigm form, can provide that anti-environmentalism indispensable to any culture seeking to avoid "complete somnambulism."[43]

In the first instance, *Dog's Life* engages its players in a typically environmental form of play. The third-person perspective is the default point of view: it is the outlook from which players start when the game is first launched, and to which they always return after having saved their progress, or following the occasional cinematic cut scenes. The virtual camera follows Jake from above, capturing his actions and immediate environment as players guide him through Clarksville, the ski resort, and beyond. This third-person perspective seems dispassionate and impartial, observing the characters and events from an abstracted, disembodied vantage point at some remove from Jake's interactions on the ground. But the game's third-person point of view is by no means detached or neutral, as the shift to first-person Smellovision retrospectively demonstrates. The explicitly subjective viewpoint of Smellovision is, as we saw, that of a dog. This canine perspective serves to remind us, however, that the ostensibly objective third-person outlook is nothing of the sort. When playing in third person, the lack of a fisheye lens means that we have a smaller field of view, the broader palette of on-screen colours allows us to experience the world as does a trichromat, and, of course, the absence of colour-coded clouds, glowing footprints, and spectacular bone-markers ensures that scents and smells are entirely unrepresented. Jake does not even take a swift, involuntary sniff as he passes through those areas that we discover, if we swap back to Smellovision, in fact contain rich deposits of appealing odours. Smellovision shows us, in short, that the default, seemingly impartial third-person perspective is, in fact, a *human* perspective.[44]

Alignment of the detached, dispassionate observer with an implicitly human mode of apprehension effectively normalizes the latter. This unacknowledged perceptual anthroponormativity functions as a McLuhanesque environment, an encompassing milieu that we fail to notice, but which nonetheless actively structures our actions and awareness. Whilst playing in third person, we do not recognize, or remember, that this apparently natural, normal point of view is but one amongst many. Indeed, when we switch to Smellovision, the zoom from an unremarked third-person perspective, which is human, to an entertainingly odd first-person perspective, which is canine, quietly normalizes the former by drawing attention only to the latter. In shifting to Smellovision, however, we are obliged at the same time to concede that the so-called third-person perspective with which we started is just as partial and particular as first. We are reminded, and compelled to confront the fact, that there is no single, standard mode of apprehension: fields of view and orientation are simply different, trichromacy is as contingent as dichromacy, and the presence or absence of detectable odours is a function of one's olfactory apparatus. Neither species-perception need be preferred to the other, or conceived as any kind of paradigm. Smellovision serves, in effect, as an anti-environment to pervasive assumptions regarding the pre-eminence of human modes of perception. By first modelling this anthroponormative outlook, and then providing a caninecentric alternative, *Dog's Life*, even whilst normalizing the human perspective, works to undermine it. Jake, the roguish outsider, welcomes us to his world, and in so doing provides a counterpoint, a counterenvironment, to

human perceptual norms. Smellovision's perceptible odours and fisheye lens draw attention to the water in which we fish swim.[45]

V working like a dog

By means of these alternate perspectives, then, the anti-environment that is *Dog's Life* invites us to appreciate something of the overlooked alterity of canine perception. Jake is the game's outsider, its alter, and, despite the necessary limitations of execution and inevitable concessions to playability, Smellovision underscores the fact that dogs and humans experience the world in radically different ways. The promise was to demonstrate *two* tricks, however, and the second, though it starts explicitly from this position of alterity, speaks at the same time to the *identity* that *Dog's Life* manages to cultivate between player and protagonist.

Building on Erving Goffman's dramaturgical analyses of social interaction,[46] in 1963 Eugene Weinstein and Paul Deutschberger published their essay "Some Dimensions of Altercasting."[47] They open with the observation that "Among the most venerable notions in social psychology is the assumption that human behavior is goal directed."[48] Consciously or otherwise, an individual will pursue these goals during the course of interactions with others, both by presenting themselves in a particular manner, and also by projecting roles or identities onto interlocutors. Such *altercasting* is a technique, Weinstein and Deutschberger argue, of interpersonal control: if an alter can be manipulated or cast into a particular role, the responses one desires of them are more likely to be elicited. Modes of altercasting can be explicit or they can guide the alter by more subtle gestures of approval and disapproval, and the literature distinguishes between manded and tact altercasts. A *manded* altercast specifies more or less directly the role that is to be adopted by the alter: they will be told unambiguously that they are a "good friend," perhaps, and thus expected to act accordingly.[49] A *tact* altercast, on the other hand, solicits a role from the alter by adopting a complementary identity. One might ostentatiously demonstrate the qualities of close friendship in the hope of evoking these in another; or, alternatively, one might adopt the role of needy neighbour in order that the alter step up to the complementary role of good Samaritan.[50] Of course, the alter may well resist these attempts to manoeuvre him or her into a particular role, whether it be submissive and helpful, dominant and decisive, or something else again. They may endeavour to engage in an altercast of their own, in fact, which is entirely at odds with the goals and objectives of their colleague. The successful continuance of the social interaction depends, then, on individuals maintaining an ongoing, negotiated working consensus, "a tacit agreement as to the roles the several participants will play out in the encounter."[51] This working consensus does not entail the roles being equal, of course, but only that at some level all parties are complicit in the interchange that results.

Weinstein and Deutschberger do not examine the question of interspecific altercasting, but it is by no means precluded by their account. The narrative that plays out during the course of *Dog's Life* illustrates a good many instances of altercasting, and although it is humans projecting the identities, insistently pursuing their goal-directed behaviours, it is the dog Jake who is altercast. Throughout the game, Jake is repeatedly cast as an itinerant, Lassie-like assistant, enjoined to complete a range of tasks and errands for polite but consistently inept humans. Submitting to entreaties or the promise of a juicy bone, Jake provides varied but characteristically canine support. In Clarksville's rustic environs, a drill sergeant turned crop farmer has difficulty protecting his seeds: "Every time I try to repair the scarecrow I get attacked by birds. Well, this means war." He addresses Jake as "Private Dog" in a direct manded altercast that effectively enlists him to scare off the crows by barking. Early in the game, Jake meets a benign, flatulent "Grandpa," Daisy's doting primary care-giver, who teaches him to sit and to lie down: "Say, Jakey, let's see what tricks you can do besides waking up a poor old man out of a lovely dream." He encourages Jake with gestures and exclamations of approval – "Sit, Jake … good dog" – and presents him with a bone when he performs correctly. The interaction is effectively a tact altercast, with Grandpa adopting the role of pedagogue whilst Jake takes up that of dutiful student. In each of these cases, and throughout the game, Jake is frequently altercast as the obedient and helpful pooch, man's best friend perhaps, a role or identity he is obliged to take on if the interaction is to persist and gameplay continue.[52]

Weinstein and Deutschberger's account of altercasting addresses a key aspect of interpersonal power relations. In his own discussions of such relations, Foucault emphasised always that, properly considered, the alter is never merely a malleable, brute object:

> a power relationship can only be articulated on the basis […] that "the other" (the one over whom power is exercised) be thoroughly recognized and maintained to the very end as a person who acts; and that […] a whole field of responses, reactions, results, and possible inventions may open up.[53]

Interpersonal control requires precisely that the alter be a person, capable of independent action, and the altercast comprises, in Foucauldian terms, a "complicated interplay" of interactions:

> In this game freedom may well appear as the condition for the exercise of power (at the same time its precondition, since freedom must exist for power to be exerted, and also its permanent support, since without the possibility of recalcitrance, power would be equivalent to a physical determination).[54]

Those occasions on which a party is subject to bare, physical determination evince the application of what Foucault designates not "power" but "capacity," a means of direct control that "stems from aptitudes directly inherent in the body or relayed by external instruments" and which is "exerted over things and gives the ability to modify, use, consume, or destroy them."[55] Although he is engaged by a succession of demanding humans, Jake, the ownerless, uninhibited canid, is never constrained or compelled by such capacities, and, within the power relations of which he is a part, a field of reactions and responses remain open to him throughout, including the option to curtail the interactions. As a canine alter, he is free, we might say, to the extent that he is at liberty to subject himself to the roles and controls of the human-initiated altercasts, or not.

Grandpa's interaction with Jake exemplifies one of the most common forms of interspecific altercast, that of training. Jake learns how to sit and to lie down, manoeuvres that will be useful to him later in the game. Techniques of the most brutal and violent kinds are undeniably employed in the course of much animal training,[56] but Grandpa is a benevolent teacher, using in his own programme only positive, linguistic reinforcement and feedback, and the final reward of a bone. There is more to the training session than simply teaching Jake these basic tricks, however. Grandpa's objective, the goal he is ultimately pursuing, is to induce Jake into seeking out his beloved Daisy. Encouraging though he is, Grandpa cannot help making unfavourable comparisons: "Nice moves, Jake, but nobody moves like my Daisy. Say, Jake, I don't suppose you could track her down? I don't know what I'd do without her." The training session is a means to an end, that of communicating Grandpa's desire and making clear the role he would like Jake to adopt: canine assistant and rescuer of the doggy damsel in distress. Interspecific altercasting, like that between humans, engenders a working consensus to which the parties more-or-less freely subscribe, but such a consensus, as Foucault argues, is the precondition for the exercise of power. The issue of interpersonal control, of attempting to realize one's goals for, or by means of, the other, lies at the heart of the animal training process, and even in those regimes where physical determination is lacking, and where direct violence plays no part, the non-human individual who is subject to the altercast will find the range of his or her actions channelled and directed. In seeking "to structure the possible field of action of others," animal training altercasters like Grandpa endeavour to shape their subject's *conduct*, where, as Foucault points out, both meanings of that equivocal term should resonate: a variety of behaviours are open to Jake, even whilst he is led in a particular direction.[57]

It is not just the human characters in *Dog's Life* who can altercast others, however. Grandpa's instruction provides just the first of a series of moves that Jake picks up during the course of his adventure. In every area of the game he meets a local dog, each of a different breed, many of whom teach him new tricks. These increasingly impressive (and unlikely) manoeuvres can be used to coax and cajole treats from nearby humans. Adopting the role of adorable performer, by sitting and begging or executing a hand stand, Jake can, without a word being spoken, tact altercast a passing human into the position of provider. Some prove immune to his charms, especially the jaded citizens of Boom City, and will send him on his way ("Get a job!"). But, provided that he is not too dirty, many will provide Jake with a choice morsel ("Ha ha, way to go there, little guy"). Interspecific altercasting can work both ways. The fact that Jake is able to assume the part of caster as well as that of alter demonstrates the intricate and fluid nature of the interplay of power relations here. To suppose that it is the caster who "holds" the power would be to fall into precisely the trap of reification, from which Foucault consistently recoiled.[58] The power *relation* between Jake and his benefactors is asymmetric in varied ways. Although Jake successfully pursues his goals by directing the behaviour of others and casting them as donors, his position is, at the same time, subservient. Not unlike the bones that Jake must collect by completing chores in order to progress through the game, the "treats" he manages to wheedle from those around him are in fact one source of the food he needs to maintain his health: they are necessary nutrition as much as indulgent titbits. Just as Jake spends a good deal of his time in the game helping others, so too he in turn finds himself dependent on the kindness of strangers.

With these interspecific altercasts, Jake demonstrates that Weinstein and Deutschberger's approach need not be confined to the human realm, and that, in its application to the give and take of interpersonal power relations, there is no reason to assume, venerably or otherwise, that it is only human behaviour that is goal directed. Further, there is an additional form of more-than-human altercasting that *Dog's Life* begs us to consider. Whenever a digital game player takes up a controller, or settles to their keyboard and mouse, or turns on their mobile device, in acquiescing to the solicitation that they press "Start," or select a difficulty level, or undertake their assigned mission, they can be considered altercast by the game. Irrespective of the genre or mode of play, gamers must yield to appropriate forms of behaviour if they wish to continue. The player's actions within the game are never entirely determined, as if controlled directly by the capacities and external instruments of the technology, of course. Rather, a whole field of responses, reactions, and possible inventions opens up, providing precisely the interaction and participation to which digital games lend themselves as a medium. The trajectory of a game, narrative or otherwise, may ultimately be as linear as any novel or film, constraining players to work their way through a succession of predetermined areas or levels. But, at the same time, players have a degree of freedom to choose how they wish to negotiate the game's challenges. The successful continuance of the interaction depends on game and player maintaining an ongoing, negotiated working consensus, a tacit agreement as to the roles that the participants will play out in the encounter.

In *Dog's Life*, then, you are altercast as a dog. The promotion and packaging take the form of an explicit, verbal cast – "It's great being a dog!" – and the gameplay itself is no less manded. Grandpa's training session does not simply teach Jake the first of his tricks but begins to instruct you, the player, as to how you must adopt this particular identity within the confines of the game. The varied forms of stereotypically canine behaviour that you must take upon yourself – digging for bones, marking territory, tracking scent trails, begging for treats, et al. – comprise, in part, the game's altercast of you in the role of a dog. Beyond these somewhat formulaic activities, though, your casting as a canid is facilitated by two technological innovations within the game. The first is Smellovision. The option to press the triangular button on the game controller and move to Smellovision is always available, and you must choose when, or even if, you will allow yourself to be so induced, in order not simply to control the avatar of Jake but to be confined to his perspective. Alexander Galloway has argued that cinematic convention most often uses the subjective shot to effect a sense of alienation, detachment or unease.[59] The first-person perspective breaks the spell of the authoritative and seemingly objective traditional camera shot, inhibiting audience identification: it is used most often to represent the vision of criminals, monsters, aliens, and "otherwise inhuman" characters.[60] Digital games that make use of the first-person perspective, on the other hand, are able to merge player and protagonist: the fact that you direct the virtual "camera" yourself ensures, in this case, a significant sense of identification.[61] With Smellovision, *Dog's Life* utilizes this mode of subjective identification but adds to it a unique, other-than-human variation. Not only is the weapon of the traditional first person shooter replaced by Jake's nose (Fig. 2), but as we saw above, when you shift from third-person perspective to Smellovision, the wide-angle view and muted tones, together with the fact that you can "see smells," imperfectly but effectively simulates canine perception. By means of this inhuman and yet entirely enjoyable identification, players are altercast as harassed but homely Jake the dog.

In addition to the distinctive visual dimension of the game, the PlayStation 2 also provides tactile feedback through the DualShock analogue controller. As Jake leaps over fences and gates, or simply bounds around the game's locations for the sheer pleasure of the experience, the controller provides an indicative jolt when he launches and lands. Similarly, whenever he strays too close to a precipitous edge, on cliffs, ski slopes or in warehouses, the controller will vibrate, cautioning the player to step back. The warning feedback on these high ledges effectively constitutes an instance of Weinstein and Deutschberger's "gestures of approval and disapproval,"[62] which serve as signposts for the route that the game would prefer you to take. Players are free to persist as they wish, but those who do so and fall will feel the heavy impact of their landing, and Jake will become noticeably less healthy as a result, complaining about his sore head and moving more slowly. Like the first-person perspective, the practice of providing indicative vibrations is by no means unique to *Dog's Life*, and is widely supported by PlayStation games. But, again, the technology here takes a unique turn. Just as the shift to Smellovision alerts us, with hindsight, to visual and olfactory aspects of the third-person perspective we had been employing, changes in our tactile experience now also prompt reassessment. Players receive haptic feedback at appropriate moments whether they are employing a first- or third-person perspective, but there is conspicuously more vibration when playing the game in subjective Smellovision than in dispassionate third-person mode. Having experienced both, the game world literally feels more physical when playing in Smellovision: it is more immediate, more "real," when you are a dog. By this means of tactile feedback, *Dog's Life* privileges a player's subjective experience of the game *as a dog* over their time merely controlling Jake from a detached but anthroponormative, third-person perspective. In addition to the visual and olfactory dimensions of Smellovision, then, this complementary form of technologically induced haptic altercasting helps affirm your identity as a dog during your interaction with the game.

VI old tricks and new

McLuhan suggests that the dog was, originally, a primordial technology that functioned as an extension of human capacities and thereby made possible a whole new way of life. There is, in McLuhan's conception of all media and technologies as "extensions of man," an anthropocentric bias which pulls against the elements of his work that, with equal force, emphasize the determining effects of the new environments generated by those technologies. In this instance, however, his equivocally humanistic probes hold the potential to highlight the means by which the technological innovations of *Dog's Life* complicate and interrogate habitual understandings of animal experience. The medium, McLuhan insisted, is not just the message, but also the massage. Technologies, he argued, produce environments that "work us over," that "leave no part of us untouched, unaffected, unaltered," a characterization that lends itself to the tactile, immersive, participative nature of digital games.[63] Games, on McLuhan's account, function as "live paradigms" of a society, encapsulating the cultural environment in model form. By simulating one situation by means of another, and inviting participation in that re-creation, games make available a mode of revealing, anti-environmental perception.[64] Participation in the particular technology at which we have been looking, in the Smellovision game mechanic that lies at the heart of *Dog's Life*, can facilitate a greater awareness of and critical reflection on traditional, clichéd ways of thinking about both animality and subjectivity, human or otherwise.

In addressing the three senses of sight, smell and touch, Smellovision provides players with an inventive and surprisingly rigorous re-creation of key aspects of canine perception as they are currently understood, even if the simulation as a whole remains necessarily simplified. The limitations of this model are, I would suggest, less important than its educative effect. During the course of his adventure, young Jake learns a number of impressive moves from his fellows, and, in turn, this new dog teaches us a pair of old tricks. On the one hand, as a ludic *altercast*, the game requires players to recognize dogs as subjects. The player of *Dog's Life* may not get to experience life as a dog (as we have seen, the game's simulated *Umwelt* and conventional narrative are inevitably schematic and stereotyped), but they are reminded that it is like *something* to be a dog. The key issue is not, *pace* McLuhan, that dogs should be conceived as technologies, augmenting human aptitudes and abilities and thereby generating new modes of social existence. Rather, this digital technology helps us to appreciate more fully the subjective dimension inherent to the interpersonal relations that pertain between the human and canine participants in any such social order, and, indeed, the asymmetric nature of those subjective interpersonal relations. The altercasts depicted and enacted by the game work to educate players, directing (but never forcing) us to acknowledge that such interspecific interactions, including training processes of all kinds, involve two more-or-less free subjects with their own perceptions and objectives who, at the same time, are engaged in an uneven, irregular, shifting power relationship.

On the other hand, as a disruptive *anti-environment*, the game requires players to reconsider the traditional, presumed primacy of the human subject. The player of *Dog's Life* does not experience events from a single, human perspective (as we have seen, our perception of the game's settings and characters is not consistent or continuous), but switches repeatedly between a human and a canine viewpoint. Contrary to first appearances, the game does not set out to provide an impersonal, impartial, third-person perspective from which locations and interactions can be observed as if with dispassionate detachment. Rather, the engaged, subjective, first-person perspective that belongs to Jake, the unruly outsider, undercuts the pre-eminence of what now turns out to have been an implicitly anthropocentric outlook. In shifting from a god's-eye to a dog's-eye view, the anti-environment established and enacted by the game works to educate players, challenging privileged, normalized notions of human perception, which is to say the perceptual anthroponormativity that would conflate an ostensibly objective with an implicitly human outlook, and reminding us that a human point of view is but one immersed, subjective perspective amongst many.[65] Thus, by means of these two edifying tricks, altercasting and anti-environments, this digital game's simulation of canine perception through its Smellovision technology complicates our understanding both of animal and of human subjects.

In terms of its narrative and characterization, *Dog's Life* may be conventional and even clichéd. Jake is both loveable rascal and faithful hound, and over the course of the game he runs the whole gamut of traditional canine conduct. Jake is not just cast but *type*cast as everyone's best friend, and a universal everydog.[66] In terms of the Smellovision technology which sets this game apart from the pack, however, *Dog's Life* mounts a genuine challenge to customary thinking about the canine. Even whilst insisting on the significant alterity of canine experience, the game obliges players to identify with the subjective dog's life of the title. Anti-environment and altercast here work together to emphasize that recognition of the otherness of different species does not preclude modes of identification. In troubling a tacit, uniform anthroponormativity, Smellovision entails a shift from an unquestioned, dogmatic identity to a disruptive, reflective alterity. At the same time, in casting players as an extraordinary, inhuman other, Smellovision involves the voluntary adoption of a projected, alien alter as an ongoing, relational identity. The anti-environmental altercast that is *Dog's Life*, which is freely taken up by its players, invites and challenges us to rethink both human and canine subjectivity. Until, that is, the novelty of this innovative, inventive digital game is exhausted, and new tricks must be sought once more.

notes

Thanks to Ewan Kirkland for introducing me to *Dog's Life*; to the Animals and Society Institute and the fellows of the inaugural Human–Animal Studies Fellowship Program at North Carolina

State University for an unparalleled research environment; to Erica Fudge, Ron Broglio, Georgina Montgomery, and Claire Molloy for providing opportunities to present my ideas to a wider audience; and to John Ensminger and Susan McHugh for their helpful comments on earlier drafts.

1 McLuhan and Watson 56–57.

2 McLuhan and Watson provide no details; on the contested question of the early collaboration, and even co-evolution, of human and canine, see Wang and Tedford esp. chapter 8; Paxton; Bradshaw 3–67 (chapters 1–2).

3 McLuhan and Watson 56–57; the idioms they quote are taken from Smith 196 (5.vi). Most of the phrases listed are in fact of very recent origin, many deriving from blood sports that bear little relation to the hunting practices in which prehistoric humans and dogs might have engaged.

4 McLuhan and Watson 57.

5 Ibid. 54.

6 *Dog's Life* was released on 29 October 2003, to mixed reviews. It was nominated for two BAFTAs ("Best Adventure Game" and "Best Children's Game"). The game should not be confused with the Macintosh "interactive storybook" published by Sanctuary Woods in 1994, titled *It's a Dog's Life* (aka *Digby the Dog* and *Digby's Adventures*).

7 *Dog's Life* booklet, p. 1.

8 Incredible as Jake's feats and adventures are, I would caution against suggesting that either he or they are "anthropomorphic"; see Tyler, "If Horses."

9 *Dog's Life* game case, reverse.

10 Uexküll 43.

11 Other attempts to represent smells within digital games have included the "cat senses" of the critically panned *Catwoman* (Electronic Arts, 2004) and the derivative "ScentView" of *WolfQuest* (Eduweb, 2007). Various devices have been developed to integrate actual smells into digital games, for instance Ruetz Scnet Systems' *Sniffman*, BIOPAC's *Scent Delivery System*, and Scent Sciences' forthcoming *ScentScape Gaming Suite*. See Steffen; Bingham; Scent Sciences.

12 An excellent overview of canine vision is provided by Miller and Murphy; on colour vision, see Nietz, Geist, and Jacobs; for accessible accounts of canine vision, see Budiansky 106–16; Horowitz 121–37; Bradshaw 224–30.

13 Miller and Murphy 1623–24, 1633.

14 Additionally, although neither is replicated in *Dog's Life*, dogs are probably more adept at motion detection, and can more easily discern fast-changing images; see Miller and Murphy 1624. The latter means that dogs have a higher threshold for flicker fusion, the point at which fast changing images blend into a constant image, which suggests that a digital game such as *Dog's Life* would appear to canine eyes as a series of distinct images, rather than as a seamless animation. Interestingly, some research has found that playing digital games actually improves visual processing (in humans); see Green and Bavelier and the literature cited therein.

15 On canine olfaction, see Ensminger; Budiansky 118–23; Horowitz 67–88; Bradshaw 230–49. On urine marking, see Scott and Fuller 69–70; in fact, "eliminative behaviour" does not seem to function as territorial marking at all, as is popularly believed and represented in *Dog's Life*.

16 In fact, several species give off a green scent, including parrots, sheep and (sometimes) pigeons. Not all creatures are visibly odorous in Smellovision, presumably in order to simplify gameplay. On cancer detection, see Horowitz 81–82; Pickel et al.

17 *Dog's Life* does not tackle the alterity of canine hearing, for instance the ability to detect ultrasonic frequencies; see Budiansky 116–18; Horowitz 92–98.

18 McLuhan and Fiore, *Medium is the Massage* 68.

19 McLuhan and Watson 57.

20 McLuhan, "Art as Anti-environment" 57. McLuhan also uses the terms *counterenvironment* (e.g., McLuhan and Watson 77; McLuhan and Parker 2; McLuhan, "Emperor's New Clothes" 342), *countersitutation* (e.g., McLuhan and Fiore, *Medium is the Massage* 68; McLuhan, "Emperor's New Clothes" 342), and *countergradient* (e.g., McLuhan and Parker 2).

21 McLuhan and Fiore, *Medium is the Massage* 88; McLuhan, "Emperor's Old Clothes" 4.

22 McLuhan, "Address at Vision 65" 226; McLuhan and Fiore, *Medium is the Massage* 93.

23 McLuhan, "Relation of Environment to Anti-environment" 112. See also idem, "Emperor's New Clothes" 342, 344–45 on Picasso.

24 McLuhan and Watson 59.

25 McLuhan and Fiore, *Medium is the Massage* 92–93; McLuhan, "Relation of Environment to Anti-environment" 112.

26 McLuhan, "Emperor's Old Clothes" 5; idem, "Address at Vision 65" 226; idem, "Relation of Environment to Anti-environment" 114; idem, "Emperor's New Clothes" 354–56.

27 McLuhan and Fiore, *Medium is the Massage* 92; McLuhan and Watson 131–33; McLuhan, *Letters* 315.

28 McLuhan, *Letters* 316; idem, "Relation of Environment to Anti-environment" 113–14.

29 McLuhan, "Relation of Environment to Anti-environment" 111–12.

30 Idem, "Classroom without Walls" 1–3.

31 Idem, "Relation of Environment to Anti-environment" 111.

32 McLuhan and Watson 198–99.

33 McLuhan, *Letters* 312.

34 Idem, "Emperor's Old Clothes" 4–5; idem, "Address at Vision 65" 226.

35 Idem, "Emperor's Old Clothes" 10.

36 In a letter to Jonathan Miller dated 8 January 1965, McLuhan pushes his anti-environmental probe still further, applying it to dreams, cosmetics, perfume, whiskers, speech, clothing, the market and prices, and more; see McLuhan, *Letters* 315. For an excellent overview and discussion of anti-environments, see Rae.

37 McLuhan, *Letters* 315; idem, "Art as Anti-environment" 57.

38 Idem, "Relation of Environment to Anti-environment" 119.

39 In 2004, Campbell's released a limited edition tomato soup with "Warhol-inspired labels." Shoppers who purchased the special four-pack could also "take advantage of an offer for a limited edition Campbell's Andy Warhol magnet set, featuring a collection of four die-cut magnets in the colorful designs of the Warhol labels." See Campbell Soup Company.

40 McLuhan, "Emperor's New Clothes" 344.

41 Idem, "Art as Anti-environment" 57.

42 McLuhan and Parker 2.

43 McLuhan and Fiore, *War and Peace* 168–69; see also McLuhan, *Understanding Media* 234–45 (chapter 24).

44 In her essay "Situated Knowledges," Donna Haraway discusses the problem of objectivity, "the god-trick of seeing everything from nowhere" (189), and the particularity and embodiment of all vision. One prompt to her own account of situated knowledges was imagining how the world looks to her dogs, which is to say "without a fovea and very few retinal cells for colour vision, but with a huge neural processing and sensory area for smells" (190).

45 On representations of the ambiguous, mixed-breed mutt as a means of social critique, see McHugh 127–70. On the ways in which another species' *Umwelt* can prompt reassessment of one's own, which he relates to the Russian Formalist notion of *ostranenie* (defamiliarization), see Winthrop-Young 230–35.

46 Goffman.

47 Weinstein and Deutschberger.

48 Ibid. 454.

49 Ibid. 456.

50 For a number of common role-pairs, see Pratkanis 216–22. The terms *manded* and *tact* come from Skinner 35–51 (chapter 3) and 81–146 (chapter 5).

51 Weinstein and Deutschberger 456.

52 This is not the only role that Jake plays within the game, however. On several occasions he becomes, instead, a mischievous mutt, sometimes at the urging of others (for instance, the children who egg the butcher in Clarkesville Centre), and sometimes on his own initiative (as when a Doberman under his control solicits a bone from a terrified shopkeeper by growling).

53 Foucault, "Subject and Power" 220.

54 Ibid. 221.

55 Ibid. 217.

56 On methods of training captive animals, see Mellen and Ellis.

57 On the term "conduct" (*conduire*), see Foucault, "Subject and Power" 220–21. Cary Wolfe has criticized Vicki Hearne precisely for her failure to reconcile the symmetry of relation that she supposes to pertain between trainer and animal, and the radical asymmetry expressed in her characterization of those animals' rights in terms of property ownership; Wolfe 44–54. In similar vein, Carol J. Adams has taken Donna Haraway to task for her defence of circus trainers, particularly her euphemistic description of performing animals as "the animals they work with," which overlooks the radically uneven nature of such a working consensus; see Adams. For further discussion of animal training as an instance of Foucauldian power relationship, see Palmer.

58 See, for instance, Foucault, "Subject and Power" 219, or idem, *Will to Knowledge* 93. On Foucault's nominalism regarding power, see Spivak 26–37.

59 Galloway.

60 Ibid. 50.

61 See Morris 89–90.

62 Weinstein and Deutschberger 456.

63 McLuhan and Fiore, *Medium is the Massage* 26.

64 McLuhan and Fiore, *War and Peace* 168–69; I have addressed the particular ways in which digital games can be considered participative elsewhere; see Tyler, "Procrustean Probe," especially the section "Retrieving Participation."

65 For illuminating discussions of literary, science fictional representations of animal otherness, see Vint.

66 It is perhaps instructive to imagine an alternative *Dog's Life* that was not motivated by the imperative of heteronormative coupling, but instead invited players to follow canine standards of sociality and sexuality.

bibliography

Adams, Carol J. "An Animal Manifesto: Gender, Identity, and Vegan-Feminism in the Twenty-First Century." *Animal Beings*. Spec. issue of *Parallax* 12.1 (2006): 120–28. Web. 30 Mar. 2012. <http://www.cyberchimp.co.uk/research/manifesto.htm>.

Bingham, Matthew. "Computer Scientists Add Smell to Games." *The Sunday Times* 26 Apr. 2009. Web. 29 July 2011. <http://technology.timesonline.co.uk/tol/news/tech_and_web/article6162217.ece>.

Bradshaw, John. *In Defence of Dogs*. London: Lane, 2011. Print.

Budiansky, Stephen. *The Truth about Dogs: An Inquiry into the Ancestry, Social Conventions, Mental Habits and Moral Fiber of Canis familiaris*. London: Phoenix, 2002. Print.

Campbell Soup Company. "Campbell's Celebrates Andy Warhol with Limited Edition Tomato Soup Cans." *PR Newswire* 14 Apr. 2004. Web. 30 Mar. 2012. <http://www.prnewswire.com/news-releases/campbells-celebrates-andy-warhol-with-limited-edition-tomato-soup-cans-giant-eagle-supermarkets-to-unveil-special-labels-and-warhol-museum-offer-72471872.html>.

Ensminger, John J. *Police and Military Dogs: Criminal Detection, Forensic Evidence, and Judicial Admissibility*. Boca Raton, FL: CRC, 2011. Print.

Foucault, Michel. "The Subject and Power." Trans. Leslie Sawyer. *Michel Foucault: Beyond Structuralism and Hermeneutics*. Ed. Hubert L. Dreyfus and Paul Rabinow. Chicago: U of Chicago P, 1982. 208–26. Print.

Foucault, Michel. *The Will to Knowledge: The History of Sexuality, Volume 1*. Trans. Robert Hurley. London: Penguin, 1998. Print.

Galloway, Alexander R. "Origins of the First Person Shooter." *Gaming: Essays on Algorithmic Culture*. Minneapolis: U of Minnesota P, 2006. 39–69. Print.

Goffman, Erving. *The Presentation of Self in Everyday Life*. Garden City: Doubleday, 1959. Print.

Green, C.S., and D. Bavelier. "Action-Video-Game Experience Alters the Spatial Resolution of Vision." *Psychological Science* 18.1 (2007): 88–94. Print.

Haraway, Donna J. "Situated Knowledges: The Science Question in Feminism and the Privilege of Partial Perspective." *Simians, Cyborgs, and Women: The Reinvention of Nature*. New York: Routledge, 1991. 183–201. Print.

Horowitz, Alexandra. *Inside of a Dog: What Dogs See, Smell and Know*. London: Simon, 2009. Print.

McHugh, Susan. *Dog*. London: Reaktion, 2004. Print.

McLuhan, Marshall. "Address at Vision 65." *Essential McLuhan*. Ed. Eric McLuhan and Frank Zingrone. New York: Basic, 1995. 219–32. Print.

McLuhan, Marshall. "Art as Anti-environment." *Art News Annual* 36 (1966): 54–57. Print.

McLuhan, Marshall. "Classroom without Walls." *Explorations in Communication*. Ed. Edmund Carpenter and Marshall McLuhan. Boston: Beacon, 1960. 1–3. Print.

McLuhan, Marshall. "The Emperor's New Clothes." 1968. *Essential McLuhan*. Ed. Eric McLuhan and Frank Zingrone. New York: Basic, 1995. 339–56. Print.

McLuhan, Marshall. "The Emperor's Old Clothes." 1966. *Unbound*, #20. Ed. Eric McLuhan and W. Terrence Gordon. Corte Madera, CA: Gingko, 2005. 3–14. Print.

McLuhan, Marshall. *Letters of Marshall McLuhan*. Ed. Matie Molinaro, Corinne McLuhan, and William Toye. Toronto: Oxford UP, 1987. Print.

McLuhan, Marshall. "The Relation of Environment to Anti-environment." 1966. *Media Research: Technology, Art, Communication*. Ed. Michel A. Moos. Amsterdam: G+B, 1997. 110–20. Print.

McLuhan, Marshall. *Understanding Media: The Extensions of Man*. 3rd printing. New York: McGraw-Hill, 1964. Print.

McLuhan, Marshall, and Quentin Fiore, with Jerome Agel. *The Medium is the Massage*. New York: Random, 1967. Print.

McLuhan, Marshall, and Quentin Fiore, with Jerome Agel. *War and Peace in the Global Village*. New York: McGraw-Hill, 1968. Print.

McLuhan, Marshall, and Harley Parker. *Through the Vanishing Point: Space in Poetry and Painting*. New York: Harper, 1968. Print.

McLuhan, Marshall, and Wilfred Watson. *From Cliché to Archetype*. New York: Viking, 1970. Print.

Mellen, Jill, and Sue Ellis. "Animal Learning and Husbandry Training." *Wild Mammals in Captivity: Principles and Techniques*. Ed. Devra G. Kleinman, Mary E. Allen, Katerina V. Thompson, and Susan Lumpkin. Chicago: U of Chicago P, 1996. 88–99. Print.

Miller, Paul E., and Christopher J. Murphy. "Vision in Dogs." *Journal of the American Veterinary Medical Association* 207.12 (1995): 1623–34. Print.

Morris, Sue. "First-Person Shooters – A Game Apparatus." *ScreenPlay: Cinema/Videogames/Interfaces*. Ed. Geoff King and Tanya Krzywinska. London: Wallflower, 2002. 81–97. Print.

Nietz, J., T. Geist, and G.H. Jacobs. "Color Vision in the Dog." *Visual Neuroscience* 3 (1989): 119–25. Print.

Palmer, Clare. "'Taming the Wild Profusion of Existing Things'? A Study of Foucault, Power and Human/Animal Relationships." *Environmental Ethics* 23.4 (2001): 339–58. Print.

Paxton, David. *Why it's OK to Talk to Your Dog: Co-evolution of People and Dogs*. Brisbane: Boolarong, 2011. Print.

Pickel, D., G.P. Manucy, D.B. Walker, S.B. Hall, and J.C. Walker. "Evidence for Canine Olfactory Detection of Melanoma." *Applied Animal Behaviour Science* 89 (2004): 107–16. Print.

Pratkanis, Anthony. "Altercasting as an Influence Tactic." *Attitudes, Behavior, and Social Context: The Role of Norms and Group Membership*. Ed. Deborah J. Terry and Michael A. Hogg. Mahwah, NJ: Erlbaum, 2000. 201–26. Print.

Rae, Alice. "Art (Anti-environment)." *Light through McLuhan*. Web. 30 Mar. 2012. <http://light-throughmcluhan.org/art.html>.

Scent Sciences. "ScentScape Gaming Suite." Web. 30 Mar. 2012. <http://www.scentsciences.com/products/scent_scape_gaming_suite.html>.

Scott, John Paul, and John L. Fuller. *Genetics and Social Behavior of the Dog*. Chicago: U of Chicago P, 1965. Print.

Skinner, B.F. *Verbal Behavior*. New York: Appleton, 1957. Print.

Smith, Logan Pearsall. *Words and Idioms: Studies in the English Language*. 5th ed. London: Constable, 1943. Print.

Spivak, Gayatri Chakravorty. "More on Power/Knowledge." *Outside in the Teaching Machine*. New York: London, 1993. 25–52. Print.

Steffen, Oliver. "High-Speed Meditation? Eine religionsästhetische und ritualtheoretische Betrachtung des Computerspiels." MA diss. U of

Bern, 30 May 2008. Web. 30 Mar. 2012. <http://www.god-mode.ch/assets/downloads/oliver-steffen_high-speed_meditation-lizentiatsarbeit%20_mai-2008.pdf>.

Tyler, Tom. "If Horses had Hands ..." *Animal Encounters*. Ed. Tom Tyler and Manuela Rossini. Leiden: Brill, 2009. 13–26. Web. 30 Mar. 2012. <http://www.cyberchimp.co.uk/research/horseshands.htm>.

Tyler, Tom. 'A Procrustean Probe.' *Game Studies* 8.2 (2008). Web. 30 Mar. 2012. <http://gamestudies.org/0802/articles/tyler>.

Uexküll, Jakob von. *A Foray into the Worlds of Animals and Humans*. Trans. Joseph D. O'Neil. Minneapolis: U of Minnesota P, 2010. Print.

Vint, Sherryl. *Animal Alterity: Science Fiction and the Question of the Animal*. Liverpool: Liverpool UP, 2010. Print.

Wang, Xiaoming, and Richard H. Tedford. *Dogs: Their Fossil Relatives and Evolutionary History*. Illustrated by Mauricio Antón. New York: Columbia UP, 2008. Print.

Weinstein, Eugene A., and Paul Deutschberger. "Some Dimensions of Altercasting." *Sociometry* 26.4 (1963): 454–66. Print.

Winthrop-Young, Geoffrey. "Afterword: Bubbles and Webs: A Backdoor Stroll through the Readings of Uexküll." *A Foray into the Worlds of Animals and Humans*. By Jakob von Uexküll. Trans. Joseph D. O'Neil. Minneapolis: U of Minnesota P, 2010. 209–43. Print.

Wolfe, Cary. *Animal Rites: American Culture, the Discourse of Species, and Posthumanist Theory*. Chicago: U of Chicago P, 2003. Print.

Ludography

Catwoman. Electronic Arts, 2004.

Dog's Life. Frontier Developments, 2003.

It's a Dog's Life aka *Digby the Dog* and *Digby's Adventures*. Sanctuary Woods, 1994.

WolfQuest. Eduweb, 2007.

book of beasts

In 1868, a slim, anonymously published book entitled *Les Chants de Maldoror* appeared in Parisian bookshops. It contained only the first *Chant* or Canto, and its purported author was the "Comte de Lautréamont" – much in line with the *fin-de-siècle* vogue for the gothic, the decadent, and a turn towards a literary, aesthetic aristocratism. In its posthumous life, the book would have a decisive impact on the counterculture, from Surrealism to Situationism. But at the time of its publication *Les Chants de Maldoror* (hereafter *Maldoror*) was completely ignored, even when Canto I was re-published in the anthology *Parfumes de l'Ame* a year later, and even when the full *Maldoror* appeared in book form.

It is not difficult to see why this book was largely ignored. It was neither prose nor poem, neither fiction nor non-fiction – it did not even fit into the popular genre of the *roman noir*, or the then-emerging genre of the prose poem. The following passage – from Canto II – is representative of much of the text's idiosyncrasies:

> There are times in life when verminous-scalped man trains his wild and staring gaze upon the green membranes of space, for ahead of him he seems to hear the ironic jeers of a phantom. He reels and bows his head: what he has heard is the voice of conscience. Then quick as a madman he rushes in amazement from the house, taking the first route available, and tears along the rugose plains of the countryside. But the yellow phantom does not lose sight of him and just as rapidly pursues. Sometimes on a stormy night while legions of winged

eugene thacker

APOPHATIC ANIMALITY
lautréamont, bachelard, and the bliss of metamorphosis

squids (at a distance resembling crows) float above the clouds and scud stiffly toward the cities of the humans, their mission to warn men to change their ways – the gloomy-eyed pebble perceives amid flashes of lightning two beings pass by, one behind the other, and, wiping away a furtive tear of compassion that trickles from its frozen eye, cries: "Certainly he deserves it; it's only justice." Having spoken thus it reverts to its timid pose and trembling nervously, continues to watch the manhunt and the vast lips of the vagina of darkness whence flow incessantly, like a river, immense shadowy spermatozoa that take flight into the dismal aether, the vast spread of their bat's wings obscuring the whole of nature and the lonely legions of squids – grown downcast viewing these ineffable and muffled fulgurations. (Lautréamont, *Maldoror* 101–02)

Animals abound in *Maldoror*, but their functions vary, from the symbolic, to the scientific, to the absurd. The passage that opens *Chant IV* extends this litany of animalization across the spectrum of beings, from the human to the mineral, culminating in one of the text's many incantations against the human:

A man or a stone or a tree is about to begin this fourth canto. When the foot slithers on a frog one feels a sensation of disgust, but one's hand has barely to stroke the human body before the skin of the fingers cracks like flakes from a block of mica being smashed by hammer blows; and even as the heart of a shark an hour dead still palpitates on the deck with dogged vitality, so are we stirred to our very depths long after the contact. Such is the horror man inspires in his own neighbour! Perhaps I am mistaken to propose this, but perhaps too I am telling the truth. I know of, conceive, a sickness more terrible than the eyes swollen from long meditations upon the strange nature of man: but I am seeking it still […] and have been unable to find it! I do not consider myself less intelligent than anyone else, and yet who would dare assert that I have succeeded in my investigations? What a lie would escape his lips! The ancient temple of Denderah lies an hour and a half away from the left bank of the Nile. Today countless phalanxes of wasps have taken possession of its gutters and cornices. They swarm round the columns like dense waves of black hair. Sole inhabitants of the cold porch, they guard entrance to antechambers as a hereditary right. I liken the humming of their metallic wings to the incessant clash of ice-floes flung against one another during the breaking-up of the polar seas. But if I ponder the conduct of him on whom providence has conferred this earth's throne, the three pinions of my grief give vent to a louder murmur! (131–32)

Much of *Maldoror* displays this sense of a prodigious outpouring of word, sound, and image – pages on pages of thick, block text with dense, phantasmagoric creatures following one after the other. At some points the text addresses the reader directly, almost confrontationally; at other points the text digresses into long, quasi-scientific descriptions of flora and fauna; at still other points the text suddenly breaks into pulp horror or gothic romance. *Maldoror* breaks nearly every rule of poetics, though it does this less as avant-garde posturing and more because the text cannot help itself. There seems to be no end to what the text can do or is capable of – in a way, its length and division into six Cantos seems almost arbitrary.

The one review that did appear upon the original publication of *Maldoror* expressed mostly a puzzled astonishment: "the hyperbolic bombast of the style, the savage strangeness, the desperate vigour of conception, the contrast of this impassioned language with the dullest lucubrations of our time, at first cast the mind into a deep amazement" (Lautréamont 276).[1] The reviewer's reaction echoes subsequent mentions of *Maldoror* in literary criticism and literary histories. Such comments have always been brief and astonished, as if happening upon a strange but menacing beast. Antonin Artaud sums up the general attitude of nineteenth- and early twentieth-century criticism towards the likes of Lautréamont, Nerval, and Baudelaire: "they were afraid that their poetry might leap out of the books and turn reality upside down" (Artaud 125). In fact, such comments on *Maldoror* are interesting for precisely this reason – they inadvertently treat the text as alive, the text as animal. For many readers, *Maldoror* not only confronted one in an abject, monstrous way, but the text itself seemed like an animal, a teratological anomaly composed of bits and pieces, a *corpus* left unfinished or untended. In contrast to the textuality of the animal so frequently found in literary representation, *Maldoror* seems to put forth the animality of the text – composed of multiple tendrils, leaping off the page, devouring the reader. In *The Romantic Agony*, published in the 1930s, literary critic Mario Praz includes only a few paragraphs on Lautréamont. But even in these few lines he echoes this tendency to treat *Maldoror* as an unnatural creature, referring to *Maldoror* as a "late but extreme case of cannibalistic Byronism" (163).

When *Maldoror* was "discovered" by the Surrealists, another shift occurred in the reception of the text: the animality of *Maldoror* suddenly gave way to an almost mystical quality. André Breton noted that *Maldoror* was "the expression of a revelation so complete it seems to exceed human potential." Philippe Soupault edited several editions of *Maldoror*, and Louis Aragon wrote the essay *Lautréamont et nous*. No doubt Breton, Soupault, and Aragon were taken by the often surprising juxtapositions in the text, the most famous of which is the phrase "beautiful as the chance meeting on a dissecting table of a sewing machine and an umbrella." But Breton's comments point to another aspect of *Maldoror*, one that extends the animality of the text, and that is its vigorous and persistent anti-humanism. While the Surrealists were unwavering critics of religion, they notoriously imported a whole host of mystical and occult themes into their works. Little surprise, then, that *Maldoror* – a text that would become part of the Surrealist canon – is frequently described in mystical terms: Breton evoked Lautréamont as "that dazzling figure of black light"; Artaud lauded *Maldoror* for its almost mystical, "perfect lucidity," an "orgy of the collective unconscious trespassing on individual consciousness"; René Daumal asserted that *Maldoror* was nothing less than a "holy war" on humanity.

These two views of *Maldoror* – as animality and as spirituality – often dovetail into each other, becoming nearly indistinguishable from each other, resulting in what Gaston Bachelard referred to as "the bliss of metamorphosis." Much of this is borne out by modern literary scholarship on *Maldoror*. There is the question of authorship – the real person Isidore Ducasse, who uses the pen name Comte de Lautréamont, and about whom very little is known, except that he was born in Montevideo, Uruguay in 1846, and that, in 1870, he died under mysterious circumstances in his Parisian hotel room. Literary detectives have noted the similarity between the name Lautréamont and

Latréaumont, a gothic novel by Eugène Sue published in the 1830s. As for the title, *Maldoror* has often been compared to the French phrase *mal d'aurore* or "evil dawn," something that seems to be supported in one of the few surviving letters of Ducasse: "Let me explain my situation to you. I have sung of evil as did Misckiéwickz, Byron, Milton, Southey, A. de Musset, Baudelaire, etc." (Lautréamont 258).[2]

Further detective work has highlighted the many textual appropriations that find their way into *Maldoror*, from literary appropriations of Homer, the Bible, Shakespeare, Dante, Baudelaire, Maturin, to extended passages from textbooks on mathematics and the natural sciences – foremost among them Jean-Charles Chenu's *Encyclopédie d'histoire naturelle* (1850–61). All of the "facts" surrounding *Maldoror* seem to point to a text that, at all levels, attempts to unhumanize itself, divesting itself of authorship, authentic voice, and even significance. The text of *Maldoror* is predatory in its extensive borrowings, and, true to Praz's words, ultimately becomes an autophage, devouring itself in the process. As we will see, this tendency towards a self-abnegation at once corporeal and textual allows *Maldoror* to effectively collapse the distance separating the bestial and the spiritual.

tooth and claw, flesh and blood

Among the modern studies of *Maldoror*, there is a significant body of work that deals with the role of animals and animality in the text. But the critical work on *Maldoror* differs on how exactly to approach the topic. Some advocate an understanding of the animals in *Maldoror* through language and linguistic tropes. For instance, in his analysis of "animal similes" in *Maldoror*, Peter Nesselroth argues that the animals in the text must be understood within the context of the cultural representations of animals, a context that allows Lautréamont to play with the many strange comparisons between humans and animals (including the many insertions of scientific descriptions of animals). The result of such disruptions in the conventions of poetic language is that the reader "finds that Lautréamont is revealing to him a new mode of perception, a vision which is not restricted by the artificial limits imposed through culture, since the boundaries between the objective and subjective have vanished" (Nesselroth 69).[3]

Other approaches suggest that the strange and anomalous creatures of *Maldoror* should be understood in terms of dream, phantasm, and the archetypes of nightmares. This is what Alex de Jonge proposes, in his analysis of *Maldoror* as an extended "anatomy of a nightmare." Animals in *Maldoror* are often portrayed in off-kilter ways, either through an inversion of the natural order (talking spiders or frogs), or through exaggerations in scale (the glow-worm as big as a house). For de Jonge, "Lautréamont distorts and destroys the essential matrices through which we decode reality. His distortions threaten our sense of space, of what is 'up' and what is 'down,' our ability to judge the relative size of images formed on the retina" (82). The particular types of animals that frequent the pages of *Maldoror* – insects, reptiles, amphibians – stand out in their radical difference from the human animal. "So foreign are they that they represent a devastating threat, a rich source of nightmare" (84).

Finally, there are those approaches that suggest we understand the innumerable animals, creatures, and monsters in *Maldoror* in relation to the concept of "nature" in the history of science, in which natural philosophy, logic and classification, and even theology go hand-in-hand. In an admirable account of animals in *Maldoror*, Alain Paris suggests that, in its excessive proliferation of life forms both real and fantastical, *Maldoror* borrows the model of the bestiary. For Paris, Lautréamont "is an explorer, an explorer of the human. An explorer of the inhuman, also, and particularly of the animal kingdom. From there the bestiary is born, like a diversion recorded in a log book" (6; translation mine).

Though one can annotate each and every instance of this or that animal in the text, such tabulations will bring one no closer to the animality of *Maldoror*. In *Maldoror*, animals are neither exemplars of the natural world, nor are they allegorical stand-ins for human beings. In a text like *Maldoror* the message is clear: *animality is no way reducible to animals*.

The critical work to have comprehended this is Gaston Bachelard's 1939 study *Lautréamont*. In his analysis of the "animal life complex" in *Maldoror*, Bachelard suggests we understand animality on a phenomenological plane, as a "vigorous poetry of aggression" (2). For Bachelard this means understanding animals less as scientific species or cultural symbols, and more in terms of their affectivity. As he notes, "Lautréamont grasps animals not as forms but as direct functions – that is, their aggressive functions" (3). Animals never are, they always do – moving, growing, pouncing, and devouring. In this sense *Maldoror* is an inventory of affects: "A complete classification of animal phobias and philias would yield a sort of *affective animal kingdom* that would be interesting to compare with the *animal kingdom* described in the bestiaries of antiquity and the Middle Ages" (79).

One of Bachelard's most instructive analyses comes in his contrast between Lautréamont and La Fontaine. The latter, well known as the author of a number of beast fables, tends to portray humans in the guise of animals. The animals in La Fontaine's *Fables* only appear to be animals; underneath they are simply exemplars of human types and characters. For Lautréamont, nearly the reverse is true – humans, when they are present in the text, tend to look like animals, or are rapidly animalized in their actions. If La Fontaine is really interested in the human in the animal, then Lautréamont is interested in the animal in the human. Furthermore, in *Maldoror* humans don't just resemble animals, they frequently undergo metamorphosis and become animals as well. Bachelard stresses this active, dynamic, "aggressive" animality in *Maldoror*, in which animality is equivalent to function. This stands in contrast to La Fontaine, for whom animality is physiognomy, representation, and form. In the innumerable hybrids, teratologies and metamorphoses that constitute *Maldoror*, Lautréamont asserts a concept of animality that is constantly producing life, life as presentation. By contrast, in the characterization and caricature of the human animals in the *Fables*, La Fontaine portrays life, life understood as representation. Whereas Lautréamont understands animality as an explosion of affects, La Fontaine understands animality as a set of behaviors. Bachelard summarizes these points, letting loose a hint of the romanticism that runs throughout his study: "La Fontaine has written of human psychology in the form of an animal fable, but Lautréamont has written an inhuman fable by reviving those brutal impulses that are still potent in men's hearts" (3).[4]

Comparisons such as these allow Bachelard to draw out his major analytical contribution to the study of *Maldoror*, and to pinpoint more specifically the animality in and of the text. A catalog of the major appearances of animals and their actions reveals for Bachelard two major aspects that constitute the animality of *Maldoror*. There is the action of tearing, and its association with the physiology of the claw, the beak, the horn, and then there is the action of sucking, and its physiological association with the sucker, the fang, the mouth. For Bachelard these constitute the twin poles of *Maldoror*'s animality: "In fact I believe Lautréamontism is almost exclusively concerned with two themes: the claw and the sucker, which correspond to the twin attractions of flesh and blood" (16).

Furthermore, the actions of sucking and tearing are not exclusive to their corresponding, anatomical organs. The action of tearing (or sucking) may be passed laterally, from a claw to a tusk, a beak, or a stinger. Similarly, the object of tearing (or sucking) may also be transferred, from the flesh of amphibian skin to the smooth marble of a statue. This lateral transference is so fecund in *Maldoror* that it almost becomes arbitrary, obtaining an almost animistic propensity for the formation and deformation of forms. As Bachelard notes, in *Maldoror* "the beautiful can no longer be simply *reproduced*. First of all it must be *produced*. It borrows from life – from matter itself – elementary energies that are first *transformed*, then *transfigured*" (60).

In one scene described by Bachelard, the disembodied "spirit" of Maldoror is transformed, first into an eagle, then into an entire flock, and finally into a strange, phantom, "lost body" composed of a detached pair of albatross wings co-mingled with a fish tail, which then takes angry flight in defiance of "the Creator." For Bachelard, such moments reveal less about animals and more about the abstract process of animalization, signaling "a sort of vertigo of the animalizing faculty, which at this point will animalize anything. In its very inadequacy this instant biological synthesis shows clearly a *need to animalize* that is at the origins of imagination" (27). Thus, while there are numerous animals in *Maldoror*, for Bachelard they are largely subsumed within the affective physiology of tearing and sucking, claw and tooth, flesh and blood.

In *Maldoror* animality is in no way identical with animals per se, and yet the text is replete with animals, animals that come to us straight out of natural history books, but also out of the fantastical worlds of literature, the bestiary, and myth. This type of layering – animals in the text and animals as text – is something that can only be achieved in the "anti-generic" poetics of text like *Maldoror*, with its many references, borrowings, appropriations, and modes of pastiche. A proposition, then: *animality is that point where the animals* in *the text and the text as animal converge*.[5] That point of convergence is, as Bachelard has already intimated, on the issue of form.

the bliss of metamorphosis

Maldoror is a text where the wild metamorphoses of creatures in the text are matched only by the equally wild metamorphosis of the text itself. So great is this animality of forms in *Maldoror* that production and destruction, generation and decay, forming and de-forming tend to overlap. As Bachelard notes, in *Maldoror* "a living being has an *appetite for forms* at least as great as his *appetite for matter*" (84). This emphasis not just on form but on life forms places Lautréamont in relation to Aristotle. While Aristotle's works in natural philosophy contain detailed descriptions of animals, it is in the treatise known as *De anima* that Aristotle talks about life in itself, apart from any particular manifestation. The question that guides Aristotle's inquiry has to do not just with understanding this or that life form but with understanding the life of every life form. For Aristotle, there must be something substantial to each and every form of life, such that we can say that a bird, a human, a tree and an octopus are all alive. It is here that Aristotle proposes the term *psukhē* (often translated in English as "soul," but more accurately as "life-principle"). This soul or life-principle is, for Aristotle, directly connected to the form of any living being. As Aristotle notes, "the soul [*psukhē*] must be substance in the sense of being the form of a natural body, which potentially has life" (69). Furthermore, this principle of life forms is always form-ing, in the sense that it is an actualization of this potential for life: "The soul may therefore be defined as the first actuality of a natural body possessing life" (69). The capacity for form is, for a thinker like Aristotle, tantamount to the potential for life; there is no life without a form of life.

Against Aristotle, form in *Maldoror* is neither that which holds matter nor that which is abstractly shaped as an empty container. Form is not an end in itself, nor does form give way to the ontological priority of matter (even if this matter is viewed in terms of its vitalistic, emergent properties). Form does not lead to that which is well-formed, or in-formed. It is, to use Bachelard's terms, the dynamics of tearing and sucking, flesh and blood, claw and tooth; but these are themselves a manifestation of a more general animality, which is driven by an aggressive, generative, "appetite for forms." Instead, in *Maldoror* there is a sense in which everything is devoured by form, at the same time that form devours everything, including itself. In *Maldoror*, form is never formed, but instead devoured, metabolized, broken down and reconstituted again in a new guise. The animality of *Maldoror* is, in a sense, the extension of Aristotelianism to its logical conclusion, in which the forming is also re-forming and de-forming as well. *Maldoror* is not exactly against Aristotelianism; if anything, it is Aristotelianism run amok, a feral Aristotelianism.

This "appetite for forms" has a teleology that is ostensibly spiritual. The appetite for forms is not arbitrary or happenstance, in spite of its aggressive, instinctual connotations. Bachelard suggests that the appetite for forms in *Maldoror* often passes through successive stages, culminating in an ecstatic, almost mystical state. The animality of *Maldoror* "would proceed through a world of living forms *executed* in well-defined bestiaries, then through a zone of *trial* forms to end finally in a more or less clear awareness of the almost anarchic freedom of spiritualization" (83). The comment is brief, and Bachelard does not follow it up. But the suggestion is an interesting one – that at the core of *Maldoror*'s animality is really a spirituality. The "spiritual anarchy" of *Maldoror* is a direct result of its aggressive, vitalistic, animality of forms. At another point in his study, Bachelard briefly mentions an even more important phrase – "the bliss of metamorphosis" – to describe this intersection of animality and mysticism: "[...] there are passages that give clear evidence of the frenzy and especially of the *bliss* of metamorphosis [...] for Lautréamont metamorphosis is a means of executing an energetic act all at once" (6). Again, Bachelard does not elaborate or develop this mystical motif. But it is arguably central to an understanding of the animality of *Maldoror*. The "bliss of metamorphosis" not only describes an animality that is inseparable from a mystical tendency but it also attempts to conceive of form in a way that is at once the lowest and the highest, the bestial and the spiritual, the deformed and the informed.

Bachelard is not the only reader of *Maldoror* to have pointed to this spiritual aspect – and in particular, the defiant, anarchic spirituality of the text. Such a reading is implicitly a part of the Surrealist fascination with Lautréamont, and it forms an important part of Maurice Blanchot's study *Lautréamont and Sade*. More recently, it has been extended in Liliane Durand-Dessert's study *La Guerre Sainte:*

Lautréamont et Isidore Ducasse, where, contra existing interpretations, Durand-Dessert argues for a "religious war" at the heart of *Maldoror*: "This revolution in consciousness of which *Maldoror* is [...] the legible trace, is not in itself a new phenomenon, since it constitutes the foundation of all religions, and is something one finds is the basis of all initiatory traditions" (3; translation mine).

In fact, we can draw out the implications of Bachelard's brief comments by re-casting *Maldoror* as a text on animality that is structured along the lines of a mystical itinerary. The text begins from a normative state of the human, structured along the division of life forms (human, animal, plant, mineral) which correspond to the division of life faculties (reason, motion, nutrition, change). We see all the exemplars of nineteenth-century European culture – the bourgeois family, the innocent youth, lovers and adventurers, men of science, and of course priests. But we also see criminals, the insane, grave-diggers, and the sick on their death-beds. In short, the life and world of humanity. In *Maldoror* this is the human world of science and religion, especially natural history and natural theology. Here, animality is always reduced to the animal, either in the form of religious iconography (e.g., sheep, goats, black dogs) or in terms of scientific rationality (e.g., the numerous appropriations of the *Encyclopédie d'histoire naturelle* found in *Maldoror*).

Into this world *Maldoror* depicts the invasion of animality, the invasion of the human by the unhuman. Sometimes this occurs via actual animals, chosen for their stark difference to the human (the various amphibians and reptile hybrids that populate the text), and at other times this occurs via animal assemblages (flocks, swarms, or packs of animals). At one point in Canto I this type of animality is depicted as a predatory, miasmatic force haunting both family and domestic space; this miasma turns into a pack of rabid dogs, which somehow telepathically connect to a frightened, young boy in the safety of his bed, as he realizes his fate. In such scenes the human characters are confronted with something radically unhuman

that they can only refer to as "evil" or "cruelty." In Canto II the discovery of a corpse in the Seine prompts a reflection on the arbitrariness of death; this directly correlates to an earlier section where the narrator imagines the corpse of God, composed of a multitude of fish and amphibians. Apprehension and dread predominate, as if the apparent order of the all-too-human world were suddenly thrown into abeyance. All human actions seem arbitrary, a pervasive sense of dread can suddenly come over the safest and most secure of situations.

Another stage occurs in which we see the animalization of the human – and the divine. Here, *Maldoror* moves beyond the confrontation of human and animal and moves into the terrain of hybridity, effected through Bachelard's two-fold activators of tearing and sucking. The relations of human–animal, animal–animal, and, importantly, animal–God are each played out through successive metamorphoses and transformations. Not only are there human–animal hybrids, we also see: a reincarnated, vampiric spider; a glow-worm's soliloquy on prostitution; a long elegy for a sleeping hermaphrodite-seer; a decapitated octopus rebelling against God; and a giant dung-beetle playing out its sorrowful, excremental, Sisyphean drama. Perhaps the most notorious example of this type of animality comes in Canto II, where, amidst a stormy shipwreck scene off a rocky shore, the narrator has a passionate sexual encounter with a giant shark.

The final stage of *Maldoror*'s mystical itinerary is the stage where the immediacy of animality leads to a pure openness of form in unceasing metamorphosis. The text departs from the animality in the text and moves out to the animality of the text itself; the scenes and events portrayed in the book – themselves appropriations from other sources – gradually give way to an almost purely arbitrary production of forms (an "anarchic spiritualization"). Vampiric fingernails grow over a few weeks, along with a craving for infant blood, there is a soliloquy by a hair found in a brothel bed, each of the body's organs turn into rodents or reptiles, God is transformed into a toad and then a giant hog, we see the world from the perspective of a python, a basilisk, and an oak tree, and mathematics becomes at once the most revered and most horrific of all things, the unhuman in its ultimate form.

Thus, insofar as *Maldoror* thinks animality through the lens of mysticism, we can discern a mystical itinerary, driven by "the bliss of metamorphosis," an animality of forms that has its telos in the dual negation of human and divine, man and God. In the final stages of this itinerary, the text moves from reproduction to production, from representation to presentation, signaling a shift from the animality in the text to the animality of the text. All of this proceeds from the premise that the animality of *Maldoror* is not reducible to animals. This in turn means a focus on the form-giving and form-generating process in life forms, taken to its extreme in the many cases of monstrous metamorphoses. Here animality is that form-generating principle of life (Aristotle's *psukhē*) that conditions the very possibility of form and forming, the very capacity for form. Philosophically speaking, such a notion implies a metaphysics of generosity, a commitment to a first principle of generation, fecundity, and affirmation – a first principle of a philosophy that, by definition, cannot be examined in philosophy. It is, perhaps, precisely for this reason that this type of animality takes on the tone of mysticism.

devouring the human

We have moved, then, from a consideration of animality in *Maldoror* to a consideration of its strange conflux of animality and spirituality, its spiritual anarchy of life forms, driven by a vitalist metaphysics of generosity and prodigality. Readers of *Maldoror* such as Bachelard already hint at this "bliss of metamorphosis," in which animality is in no way reducible to animals.

But in our shift from animality to spirituality we have, as it were, jumped over the human – the target of some of the most vitriolic phrases in *Maldoror*. What is the status of the human in relation to this mystical animality, this bliss

of metamorphosis? Bachelard gives us a clue, though it is ultimately the weakest point in his analysis. There is undoubtedly a strong anti-humanist thread that runs through *Maldoror*, evidenced not only by the violence done to nearly all human characters but also by the bestiary of animals, animal parts, and animal transformations. While human beings usually undergo monstrous transformations in *Maldoror*, and while the author constantly rallies against the human (including the human-made God), Bachelard suggests that the anti-humanism of *Maldoror* is actually an attempt to re-imagine a renewed human, within the framework of animality. As Bachelard notes provocatively, "the bestiary of our dreams animates a life that returns to biological depths [...] All the functions can create symbols; all biological heresies can produce phantasms" (86).

Ultimately, for Bachelard the animality of *Maldoror* comes to be regarded as a testament to a kind of super-human capacity for imagination and the poetics of form – even at the moment when the text so violently revolts against the human. For Bachelard the animality of *Maldoror* is really about *poiesis*, the creation of new forms, what Bachelard terms a "poetry of the project" or simply the "open imagination." The successive, incessant, and aggressive transformations in *Maldoror* come to be seen as an indicator of the human passing beyond the animal, and finally passing beyond itself. Bachelard again: "At that point, man appears as the sum of vital possibilities, as a *super-animal*. All of animality is at his disposal" (11).

Bachelard's emphasis on the imagination is not an attempt to re-install the human at the top of the Great Chain of Being. Lurking behind his proposals for a *poiesis* of the open imagination is a further investment in a vitalist metaphysics of fecundity, generosity, and metamorphosis. *Maldoror* is not so much a testament to the human capacity to imagine any- and everything; rather, *Maldoror* is the manifestation of an anonymous, generative, creative principle that animates the world – and especially the natural world. There is a sense in which, for Bachelard, "imagination" is simply another name for *poiesis*, producing, creating. This comes through in the closing pages of his book:

> The way of direct human exertion is only a poor extension of animal exertion. It is in the *dream of action* that the truly human joy of action resides. To cause to act without acting; to leave bound time for liberated time, performance time for decision time [...] to replace a philosophy of action, too often a philosophy of agitation, with a philosophy of repose and then a philosophy of the consciousness of reposing [...] Next, this repose of the imagination must be taken as the point of departure for a discovery of firmly deanimalized thought-motifs, free from all allurement, cut off from the hypnotism of images, clearly separated from the *categories* of understanding that are concretized forms of intellectual prudence, "fossilized states of intellectual repression." (90)

While we may agree with Bachelard's attempts to move animality beyond "the animal," it remains unclear how such a "firmly deanimalized thought" avoids simply re-instating the human at the top of the pyramid (all the more so in that Bachelard mentions, at one point, a "psychoanalysis of life" as part of his project). Part of the confusion comes from the jump that Bachelard makes, from a discussion of the animality in the text to a discussion of the animality of the text of *Maldoror* itself. While much of Bachelard's analysis operates at the level of the animality in the text (as per his analysis of tearing and sucking motifs), the larger, more philosophical claims made by Bachelard rely on an identity between the two types of animality (in the text and of the text). Put another way, Bachelard's arguments concerning the open imagination and the vitalist poetics of *Maldoror* actually rely on the collapse of the distinction between the animality in the text and the animality of the text.

In his discussion of the open imagination, Bachelard wants to make a claim for a generalized *poiesis*, one that extends beyond modes of literary representation, and, ultimately, even beyond modes of textual production. This *poiesis* is, for Bachelard, thoroughly vitalist, generative, and fecund; it is an animality of

perpetual transformation and metamorphosis. In this way, the open imagination is not exclusive to human beings or the products of human culture. Bachelard comes very close to making a "realist" assertion for this generalized *poiesis* as a property of the world as such.

Yet, even if we take Bachelard to mean "imagination" in a broadly non-human sense, it is unclear why an explicitly poetic text, belonging explicitly to human culture, would be the avatar of this open imagination. Why not, for instance, make the claim that the generative metamorphoses and morphologies of nature itself are the true exemplars of the open imagination? Problems like these are compounded when one considers the strong, anti-humanist thread that runs throughout *Maldoror*. For a text that so virulently poises itself against humanity, it would seem strange to hold up *Maldoror* as the pinnacle of imaginative achievement, human or otherwise.

For Bachelard, anti-humanism and animality co-exist in an uneasy relationship. Anti-humanism ultimately leads to a renewed humanism of the imagination, an elevation of the human above the animal. Hence Bachelard manages to take what is perhaps the most extreme statement of anti-humanism and transform it into a call for a renewed humanism of the imagination. For Bachelard, the animality of *Maldoror* is the vehicle by which humanity renews itself – less in terms of reason and more in terms of imagination. Hence animality is subsumed within humanity (though a humanity of imagination rather than reason). *Maldoror* is, in Bachelard's reading, a poetic *sacrifice* of animality for a new humanity.

apophatic animality

As we have noted, *Maldoror* suggests to us a relationship between animality and spirituality. But this spirituality need not be of an anthropocentric type, in which the human is the endpoint and culmination of animality, immediacy the highest bliss, metamorphosis the highest form. Might there be another type of animality at play in *Maldoror*, an animality that still retains this link to spirituality, but that does not simply culminate in the raising up of the human animal as its highest form? Is there an anti-humanist animality at play in *Maldoror*?

The traditional view of animals in philosophy is often split: it tends either to reduce animals to their naturalistic substrate or elevate them into an abstract realm of flows and forces. Ironically, in the philosophy of animals, animals themselves disappear behind the mists of empirical observation, epistemological classification, and the hermeneutic demand for myth, symbol, and psychological depth. Either animality is reducible to animals or animality is raised up to and includes human life; either animality excludes the human or it includes the human. In the former view – a lateral version of animality – all life forms are arrayed on a plane as part of the animal kingdom, each with differing characteristics. In the latter view – a vertical version of animality – all life forms are arranged in a hierarchy of capacities and functions. However, what both of these views have in common is a philosophical commitment to a *metaphysics of generosity* and prodigality, a vitalist ontology of fecund forms that constantly proliferate, generate, and change. The animality of animals is this commitment to the idea of a primordial, vitalist generosity of creation and form. Hence the philosophy of animals must presume something prior to the animal called "life," that is connected, in some basic way, to the generation and proliferation of forms that constitutes this traditionalist view of animality. This metaphysics of generosity is, in short, the a priori of animality.

At the crux of the bliss of metamorphosis and its mystical itinerary there is a philosophical commitment to a notion of life as generative, fecund, and proliferating of life forms. There is a metaphysics of generosity that determines and conditions the bliss of metamorphosis, and it is this vitalist metaphysics that also plays into the glorification of the human in terms of its creative capacity and the "open imagination." In short, everything is generous, and it is generous for the benefit of the human.

If this is the case, then we can ask whether there is in *Maldoror* an animality that is less a vitalistic, generous, bliss of metamorphosis,

and more its inverse: an animality of the negation of life, of the withdrawal of form, of the liquidation and dissipation of form, of the emptying of all form – in short, an animality that, while retaining the link to mysticism, also conceives of mysticism precisely as the dissipation of form. Is there an animality of absence, of distance, of opacity – an oblique animality? This would mean looking to those moments in *Maldoror* when animality ceases simply to be immediate, while not being absent – moments when animality avails itself in its inaccessibility. Animality in *Maldoror* is not the continual proliferation of forms; animality constantly slips away, a single animal losing its form and becoming a swarm of animals, which in turn disaggregate and become indistinguishable from the elements and the atmosphere itself, all of this just as easily dissipated into the oblique and opaque *ideas* that inhabit the text of *Maldoror*.

Perhaps, then, there is less a mystical itinerary in *Maldoror* and more a mystical anomaly, an interruption of divine beatitude, a mysticism that ends not with the fullness of the bliss of metamorphosis but with a different type of bliss, a gothic bliss of the loss of all form. Such a gothic bliss would require an ontology based not in affirmation and generosity, but instead in negation and dissipation – in short, it would require a negative theology. In this sense, *Maldoror* is less a vigorous bliss of metamorphosis, and more an incessant poetry of negation. The fifth to sixth-century mystic Dionysius the Areopagite articulates this contrast between affirmative (*cataphatic*) and negative (*apophatic*) forms of mysticism:

> Now it seems to me that we should praise the denials quite differently than we do the assertions. When we made assertions we began with the first things, moved down through intermediate terms until we reached the last things. But now as we climb from the last things up to the most primary we deny all things so that we may unhiddenly know that unknowing which itself is hidden from all those possessed of knowing amid beings, so that we may see above being that darkness concealed from all the light among beings. (138)

The negation of this type of mysticism is not a negation of privation or subtraction but instead a contradictory negation that is actually superlative, precisely because it forms the horizon of human knowledge. It is a negation that involves an erasure, an effacement, a denial (*apo-*) of rational discourse and thinking (*phanai*). Borrowing from the apophatic tradition of negation in mysticism, we might call this an *apophatic animality*.

In *Maldoror*, apophatic animality has two aspects, both of which have to do with the negation of form. On the one hand there is *anamorphosis*, exemplified by the many chimeras, monsters, and hybrids that populate *Maldoror*.[6] A passage from Canto IV illustrates the dual process of building up and breaking down of form that is part of the apophaticism of *Maldoror*:

> I am filthy. Lice gnaw me. Swine, when they look at me, vomit. The scabs and sores of leprosy have scaled my skin, which is coated with yellowish pus. I know not river water nor the clouds' dew. From my nape, as from a dungheap, sprouts an enormous toadstool with unbelliferous peduncles. Seated on a shapeless chunk of furniture, I have not moved a limb for four centuries. My feet have taken root in the soil forming a sort of perennial vegetation – not yet quite plant-life though no longer flesh – as far as my belly, and filled with vile parasites. My heart, however, is still beating. But how could it beat if the decay and effluvia of my carcass (I dare not say body) did not abundantly feed it? In my left armpit a family of toads has taken up residence, and whenever one of them moves it tickles me. Take care lest one escape and come scratching with its mouth at the interior of your ear: it could next penetrate into your brain. In my right armpit there is a chameleon which endlessly chases the toads so as not to die of hunger: everyone has to live. But when one side completely foils the tricks of the other, they like nothing better than to make themselves at home and suck the dainty grease that covers my sides: I am used to it. A spiteful viper has devoured my prick and taken its place. This villain made a eunuch of me. Oh! If

> only I could have defended myself with my paralysed arms – but I rather think they have turned into logs. Be that as it may, it is vital to note that in them blood no longer pulses redly. Two small hedgehogs, that grow no more, have flung to a dog – which did not decline them – the contents of my testicles; inside the scrupulously scrubbed scrotal sac they lodged. My anus has been blocked by a crab. Encouraged by my inertia, it guards the entrance with its pincers and cause me considerable pain! Two jellyfish crossed the seas, at once enticed by a hope which did not prove mistaken. They closely inspected the two plump portions which comprise the human rump and, fastening on to these convex contours, so squashed them by constant pressure that the two lumps of flesh disappeared while the two monsters which issued from the kingdom of viscosity remained, alike in colour, form, and ferocity. Speak not of my spinal column, since it is a sword. (142–43)

In this non-narrative section we are given a monstrous version not only of the body natural but the body politic as well. The body depicted is at once decaying and, one senses, about to crumble and fall apart – and yet it is also fixed, frozen, and petrified in its place. In anamorphosis one sees the breakdown of part–whole relationships, in favor of the play between part and part, but also between whole and whole. Anamorphosis can take place in space – as in the above citation – or it can take place in time, as in the final Canto, where an archangel is turned into a giant edible crab, then into a fishtail with bird wings, and so forth. Anamorphosis functions on the axis of humanity/animality; its operator is that of decay and decomposition.

In addition to anamorphosis, there is also *amorphosis*, exemplified by the numerous instances of formlessness in *Maldoror*. Here, the animality of formlessness need not have to do with actual animals. Whereas anamorphosis is predominantly metamorphic, amorphosis is predominantly morphological, dealing with the limits of form and formlessness. An example is given in Canto V of *Maldoror*, which, interestingly, becomes a meditation on the poetics of the text itself:

> Let not the reader lose his temper with me if my prose has not the felicity to please him. You maintain my ideas are at least singular. What you say, respectable man, is the truth, but a half-truth. And what an abundant source of errors and misapprehensions every half-truth is! Flights of starlings have a way of flying which is theirs alone and seems governed by uniform and regular tactics as a disciplined regiment would be, obeying a single leader's voice with precision. The starlings obey the voice of instinct, and their instinct leads them to bunch into the centre of the squad, while the speed of their flight bears them constantly beyond it; so that this multitude of birds thus united by a common tendency towards the same magnetic point, unceasingly coming and going, circulating and crisscrossing in all directions, forms a sort of highly agitated whirlpool whose whole mass, without following a fixed course seems to have a general wheeling movement round itself resulting from the particular circulatory motions appropriate to each of its parts, and whose centre, perpetually tending to expand but continually compressed, pushed back by the contrary stress of the surrounding lines bearing upon it, is constantly denser than any of these lines, which are themselves the denser the nearer they are to the centre. Despite this strange way of swirling, the starlings cleave through the ambient air at no less rare a speed and each second make precious, appreciable headway towards the end of their hardships and the goal of their pilgrimage. Likewise, reader, pay no attention to the bizarre way in which I sing each of these stanzas. (159–60)

Unlike other instances of animality in *Maldoror*, with their incessant biological admixtures and hybrids, here we have a sustained passage on a single phenomenon, that of swarming behavior that is at once tightly organized and yet formless and chaotic. It is also a passage that is itself borrowed from the *Encyclopédie d'histoire naturelle*, making the very act of reading the text ambiguous in its conflation of scientific description and figurative simile.[7] In amorphosis form is pushed to its limit, becoming either the absence of all form (the evacuation of all form) or absolute form (the devouring of all possible form). In *Maldoror*, these instances of formlessness can exist within a single body (as in the morphologies of the Maldoror character as a pack of dogs and then a miasma), or it can exist pervasively throughout multiple bodies (e.g. flocks of birds, a horde of rats, a swarm of flying squids). Amorphosis functions along the axis of humanity/divinity; its operator is that of dissipation and dissolution.

Maldoror is an anomalous text replete with animals of all kinds. It is also a text that is equally concerned with animality, an animality not reducible to animals, an assertion made in the text's frequent transgression of both naturalistic and narrative form. But the question is whether a text like *Maldoror* is about the production of forms or the loss of form, whether the text points to an animality that is driven by a vitalist generosity of life or by an apophatic dissipation of life.

Hence we have two variants on a theme, a theme concerning the intersection of animality and spirituality. On the one hand we have the bliss of metamorphosis. This relies on a philosophical premise, in which a metaphysics of vitalist generosity provides the backdrop for a fecund, proliferating, creation of life forms – a bliss of metamorphosis that is ultimately recognized to be as spiritual as it is animal. With the bliss of metamorphosis, we have both a lateral animality and vertical humanity, the former by virtue of the many life forms presented in the text, which reach their highest pitch in a humanized capacity for *poiesis*. For the bliss of metamorphosis, animality is immediate and affirmative of life.

By contrast, we have presented another variation, which highlights the unavoidable anti-humanism that energetically drives *Maldoror's* litany of assaults against both God and man. In this view, *Maldoror* is a text that rails against the human – and also against "life" (insofar as life is the privileged designation made by humans on behalf of other beings). While Lautréamont takes up the Aristotelian fascination with life forms, he is also positioned against Aristotle on the need for a life-principle and the metaphysical necessity of its ability to cause form to take shape. With Lautréamont, we move from the well-formed life (the life of, say, biological classification), through the form-ing life (the metaphysics of generosity, the bliss of metamorphosis), to a more suspect and shadowy region, a gothic mode where form and forming is inseparable from de-forming and un-forming. If it is still "mystical," it is in this apophatic register, summarized in Alain Paris' study:

> Lautréamont deplores the *human* form of conscience, which is duality and conscious-ness of this duality – that is to say, the con-sciousness of the separation of the self and the world, and of the self with itself [...] For Lautréamont, God frequently represents this alterity that both founds consciousness and the tearing of consciousness from itself [...] It is in this way that one must under-stand the hatred of God, invoked from the beginning of *Maldoror*, and not in the tra-ditional sense of the problem of good and evil [...] There is a mysticism of hatred in Lautréamont. Hatred is a propaedeutic of the divine and that which is beyond the human. (115–16; translation mine)

In Bachelard's reading, the animality of *Mal-doror* lies in the "bliss of metamorphosis," a concept of animality that is a conjunction of the immediacy of life with a technics of form. In his evocations of the "open imagination," Bachelard therefore intimates a reading of *Maldoror* as a heroic type of poetry, an example of the modernist imperative to discover the new.

However, this tends to downplay the central importance of the gothic in *Maldoror*, both in its style and in its literary context. In this gothic mode, life exists only to the extent that it constantly ceases to exist; the prodigality of forms only exists in so far as they are decaying, decomposing, or disintegrating. In the gothic mode, animality is a form of life that grows by decaying, that is built in ruins, and that is prodigious in its nothingness. In short, *Maldoror* is less a *heroic* and more a *tragic* type of poetry.

Maldoror is a tragic type of poetry because it asserts that there is too much form in the world. This is because there is too much life (and there is no form without life). *Maldoror* attempts an impossible task, which is to actively and conti-nually un-form all form, above all that most tiring of forms, the human form. In spite of its many invectives against God, and in spite of its many absurdist descriptions of animals, the challenge posed by *Maldoror* is not a chal-lenge against religion or science. The real chal-lenge posed by *Maldoror* is this: *what is the most adequate form of the unhuman?* And yet *Maldoror* can only accomplish this via some form; hence its poetics of gothic misanthropy must take on the abandoned shell or the carcass of existing forms, both of literature and of life.

coda: non-literature and non-life

Near the end of his study, Bachelard asks how a text such as *Maldoror* might impact not only literature or poetry but the entire field of poetics itself. It leads him to coin a somewhat cumbersome yet evocative term, "non-Lautréamontism":

> Ducasse's metamorphoses have had the advantage of un-anchoring a type of poetry submerged in the job of describing. In my opinion we must now take advantage of a life given over to the metamorphosing powers in order to move on to a sort of *non-Lautréamontism* that will spill out of *Maldoror* in all directions. I shall continue to use the term "non-Lautréamontism" while giving it the same function as that non-Euclidianism which can generalize Eucli-dean geometry. (90)

Bachelard's evocation of a non-Lautréamontism looks forward to the "non-philosophy" of Fran-çois Laruelle, who also makes the comparison to non-Euclidian geometry. For Laruelle, non-philosophy is neither anti-philosophy nor meta-philosophy. It takes philosophy as its raw material, illuminating the "philosophical decision" that structures the separation of phil-osophy from theology, mathematics, or poetry, and that also internally distinguishes fundamental philosophy (metaphysics and ontology) from regional philosophy (the philosophy of religion, political philosophy, the philosophy of science). This philosophical decision is philosophy's necessary self-positing and the basis of its explanatory power. As Laruelle asserts, "philosophy is regulated in accordance with a principle higher than that of Reason: the *Principle of sufficient philosophy*. The latter expresses philosophy's absolute autonomy, its essence as *self*-positing/ donating/naming/deciding/grounding" (Laruelle 139). A non-philosophy would examine those aspects of philosophy that philosophy itself cannot examine, without its becoming something else (a logic, a science, a poetics).

If we interpret Bachelard's proposal for a non-Lautréamontism or a non-literature in this way then the question would be whether *Maldoror* is a work of non-literature, in the sense that it complicates a poetics of balanced form and content, a literature of representation and hermeneutic depth. This returns us to our opening comments about animality in *Maldoror* – the animality in the text and the text as animality. *Maldoror* is non-literature because in every phrase it questions the "literary decision" that literature be at once apart from and yet engaged in that which it depicts. *Maldoror* takes aim at the human per se, but also at the cultural concept of the human as a literary creature, a form of life given over to reflection, representation, and the production of meaning.

But as Gilles Deleuze reminds us, literature is indelibly linked to life. However, this need not be in the usual sense, in which literature represents life, as form for matter. As Deleuze asserts, "[t]o write is certainly not to impose a form (of expression) on the matter of lived experience" (225). This is because literature not only transforms, but is transformed as well: "Literature rather moves in the direction of the ill-formed or the incomplete [...] It is a passage of Life that traverses both the livable and the lived" (ibid.). If this is the case, if literature and life are connected not as form to matter but as mutually deforming and unforming activities, then what is the corollary for a "non-Lautréamontism" or a "non-literature"?

It would seem that any non-Lautréamontism or non-literature immediately raises the possibility of a *non-life*, a life that cannot be lived, or, better yet, the "lived-without-life."[8] Bachelard hints at this: "*Maldoror* can be taken as a pretext for understanding what a work would be if it were somehow to tear away from ordinary existence and welcome that other life which must be designated by a contradictory neologism as an *unlivable life*" (55). Like Deleuze, Bachelard also argues for a non-representational notion of literature, a notion of literature that is itself a manifestation of something immanent to both literature and life. But this also pulls both thinkers towards a metaphysics of generosity, a vitalist commitment to dynamic change and constant becoming.

Be that as it may, this relationship between literature and life is only given testament within the exclusive provenance of humanity. To whom is literature and its relation to life directed, if not to the specifically human life that is able to qualify both life and literature? This is one of the central challenges put forth by *Maldoror*. The strange proposition of a non-literature, a literature not intended for humans, would seem also to necessitate a non-life, or a life that is neither simply human life-experience nor that of the life sciences. *Maldoror* is in every way a text poised against the human, even in terms of its literary form:

> Here in fact is a work born not from the observation of others but not exactly from the observation of oneself either. Before being observed it was created. It has no goal yet is an action. It has no plan yet is coherent. Its language is not the expression of a previous thought. It is the expression of a psychic force that has suddenly become a language. In short it is instant language. (Bachelard 55)

But Bachelard's study is ultimately compromised by its "heroic" turn towards the human. Not only does Bachelard recuperate animality within a higher humanity (the "open imagination" with all of the animality at its disposal) but nearly all of Bachelard's discussion on animality and the "bliss of metamorphosis" centers on the almost instinctual *immediacy* of animality – aggression, action, movement, creation, spontaneity, and so on. For Bachelard, animality is equivalent to immediacy. What is more, Bachelard emphasizes the fact that *Maldoror* seems to itself – as a text – be equivalent to this immediacy, as if the text itself spontaneously preceded its own conceptualization, planning, and writing. Constantly changing, continually forming hybrids, incessantly pouring forth pages and pages of "the bliss of metamorphosis," *Maldoror* seems to have always existed, prior to any and all forms of mediation. The animality of *Maldoror* puts forth what is really a philosophical challenge – it is an immediacy that challenges all philosophical decision, any metaphysical claim that would begin by, for instance, dividing being from non-being, thought from world, the living from the non-living, life from death. Everything is possible in *Maldoror*, all hybrids are permitted, all forms only exist to be deformed and reformed. Teratology in fact becomes the norm, with its propensity for the aggressive, spontaneous creation of novel forms. In this sense *Maldoror* also poses a challenge to the principle of sufficient reason, a moral and theological principle that the world is well formed, and that the form of the world is necessary to the world.

notes

1 The review was published in *La Jeunesse* 5 (Sept. 1868), and is reprinted in *Maldoror: The Complete Works of the Comte de Lautréamont*.

2 From a letter to Poulet-Malassis, 23 Oct. 1869, reprinted in *Maldoror: The Complete Works of the Comte de Lautréamont*.

3 This is also the approach adopted by works associated with the journal *Tel Quel*, the most notable example of which is Julia Kristeva's *Revolution in Poetic Language*.

4 Bachelard also goes on to contrast *Maldoror* to Kafka's "Metamorphosis," Kipling's *The Jungle Book*, and Wells's *The Island of Dr. Moreau*.

5 Lautréamont is not the first to suggest this intersection – arguably it can be found in early modern bestiaries and teratologies, where the language of science and fable intermingle. More importantly, it is a point taken up in the contemporary theory, in Deleuze and Guattari's discussion of animality in *Kafka*, and in Akira Lippit's notion of "animetaphor," developed in his book *Electric Animal*.

6 Although the term anamorphosis has a double meaning – in art history to describe a visual illusion, and in biology to describe the development of embryonic life forms – I am using the term to describe a breaking-down of form and the forming capacity. Thus *ana-morphosis* is, in this case, a literal layering of negative form (*ana*- "back," "reversion," "again") on top of existing form (*morphē*, "shape," "form").

7 The appropriation of this passage is pointed out in a still-useful 1952 article by Maurice Viroux, "Lautréamont et le Dr. Chenu" (published in the *Mercure de France*), where Viroux traces it to an almost verbatim passage in the volume *Oiseaux* of Chenu's *Encyclopédie*.

8 I borrow this phrase, with slight changes, from Laruelle. In his book *Mystique non-philosophique à l'usage des contemporains*, Laruelle discusses the immanent type of mysticism represented in the works of Eckhart, described as a "Vécu-sans-Vie" (Lived-without-Life).

bibliography

Aristotle. *On the Soul/Parva Naturalia/On Breath*. Trans. W.S. Hett. Cambridge, MA: Harvard UP/ Loeb Classical Library, 2000. Print.

Artaud, Antonin. "Letter on Lautréamont." *Artaud Anthology*. San Francisco: City Lights, 1965. 123–27. Print.

Bachelard, Gaston. *Lautréamont*. 1939. Trans. James Hillman and Robert S. Dupree. Dallas: Dallas Institute, 1986. Print.

de Jonge, Alex. *Nightmare Culture: Lautréamont and "Les Chants de Maldoror."* New York: St. Martin's, 1973. Print.

Deleuze, Gilles. "Literature and Life." Trans. Daniel Smith. *Critical Inquiry* 23.2 (1997): 225–30. Print.

Dionysius the Areopagite. *Pseudo-Dionysius: The Complete Works*. Trans. Paul Rorem. New York: Paulist, 1988. Print.

Durand-Dessert, Liliane. *La Guerre Sainte: Lautréamont et Isidore Ducasse*. Nancy: Presses Universitaires de Nancy, 1991. Print.

Laruelle, François. "A Summary of Non-philosophy." Trans. Ray Brassier. *Pli: The Warwick Journal of Philosophy* 8 (1999): 138–48. Print.

Lautréamont, Comte de. *Maldoror: The Complete Works of the Comte de Lautréamont*. Trans. Alexis Lykiard. Cambridge: Exact Change, 1994. Print.

Nesselroth, Peter. *Lautréamont's Imagery: A Stylistic Approach*. Paris: Droz, 1969. Print.

Paris, Alain. "Le Bestiare des Chants de Maldoror." *Quatre Lectures de Lautréamont*. Paris: Nizet, 1972. 83–143. Print.

Praz, Mario. *The Romantic Agony*. 1933. Trans. Angus Davidson. New York: Meridian, 1963. Print.

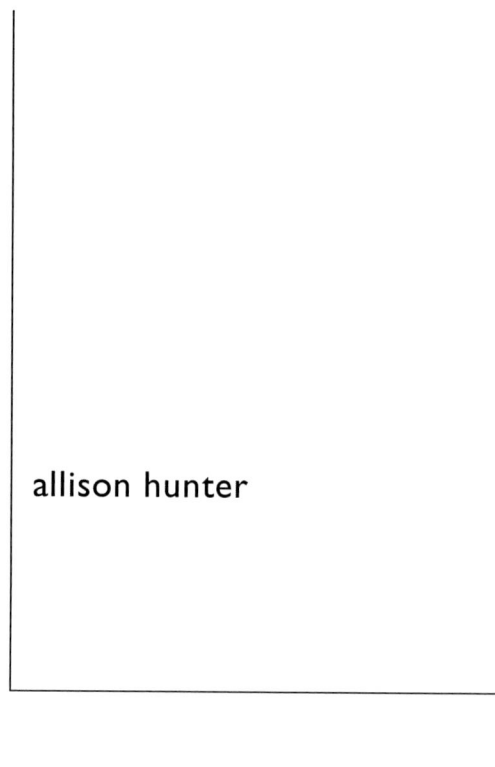

allison hunter

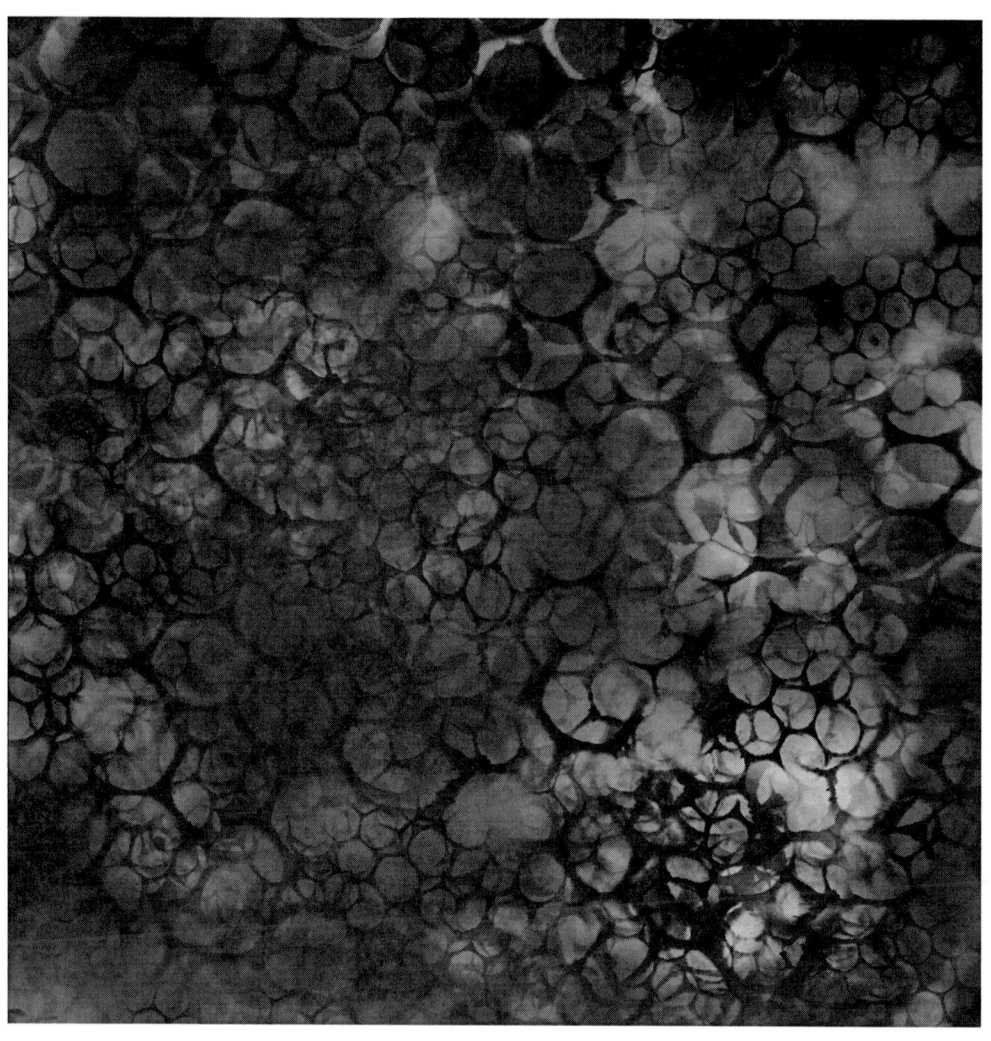

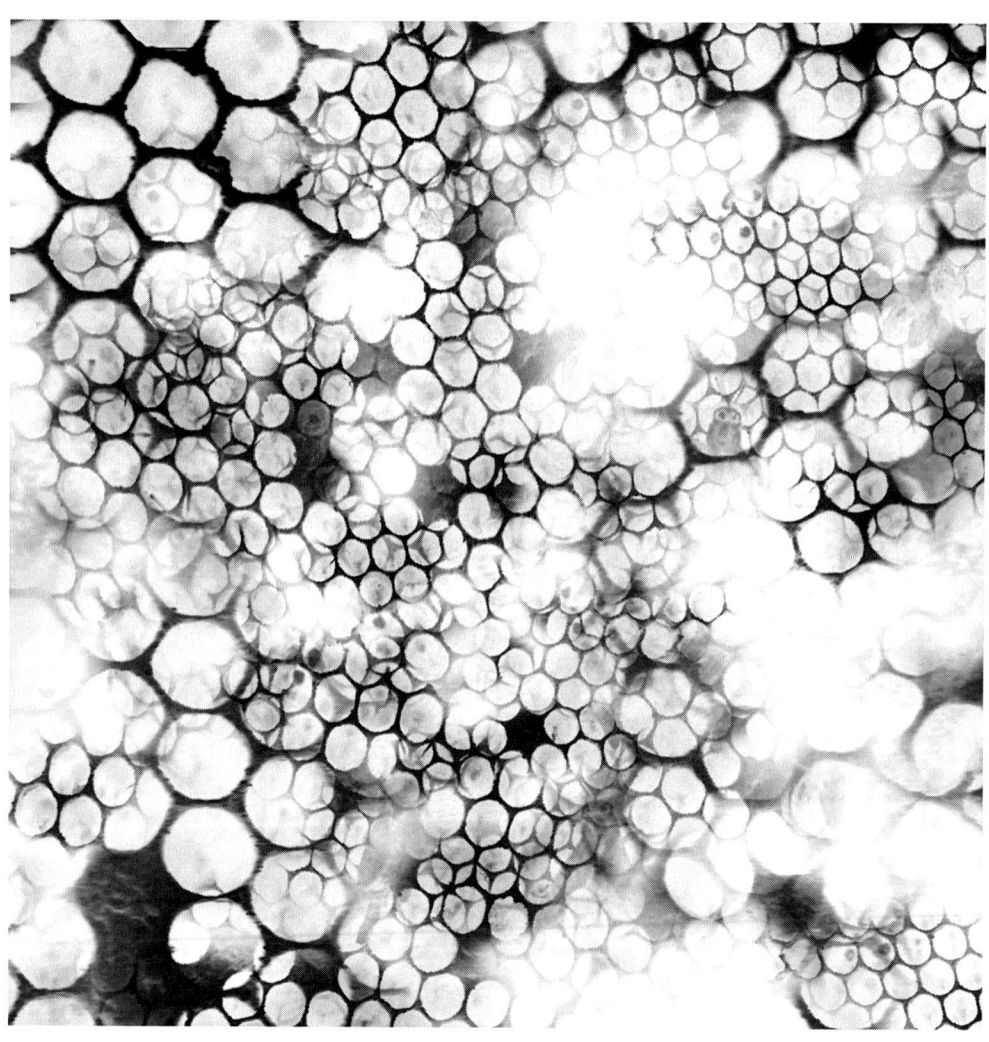

images

1. Allison Hunter, "Untitled (dragonfly)," 2009, 33 × 50 inches, digital c-print.
2. Allison Hunter, "Untitled #1 (from the Honeycomb series)," 2011, 30 × 30 inches, digital c-print.
3. Allison Hunter, "Untitled #2 (from the Honeycomb series)," 2011, 30 × 30 inches, digital c-print.
4. Allison Hunter, "Untitled #3 (from the Honeycomb series)," 2011, 30 × 30 inches, digital c-print.
5. Allison Hunter, "Untitled #4 (from the Honeycomb series)," 2011, 30 × 30 inches, digital c-print.
6. Allison Hunter, "Untitled #5 (from the Honeycomb series)," 2011, 30 × 30 inches, digital c-print.

I

I will end with an emphasis on the invisible, but start with disappearance. The two ends connect in this text, which concerns the intertwining of the ecological with the aesthetic; not an aesthetics *of* nature, or ecology, or even ecocrisis, but how such are themselves perceptible through an intertwining with technological epistemologies. Indeed, this deals with capacities of bodies stretched across various ecologies in the manner Guattari talked about, and what recent techno-aesthetic theory concerning sensation has been arguing (Parisi). In short, what such a stance is saying is that the capacities of human and animal bodies cannot be detached from considerations of their technological framings, which in this text is a question of ecology – a feedback loop of various levels and scales. In this sense, this text focuses on how to think the visual culture of disappearance – more closely, disappearance of animals.

Hence, by way of a preface, let's start with Ernst Jünger's novel *The Glass Bees* from 1957 – a science fiction story of an automata producer and industrialist Zapparoni and his miniature robotics that, according to Bruce Sterling's introduction to the book, resemble more the high-tech creatures of recent years of MIT design laboratories than the clunky robotics of typical 1950s science fiction. Indeed, Sterling's characterization echoes the German media theorist Friedrich Kittler, when the former writes of *The Glass Bees* and Jünger of how he "understands that technology is pursued not to accelerate progress but to intensify power" (x). If Jünger's earlier novel *The Storm of Steel* became a key reference point for a certain brand of (German) media theory that

jussi parikka

INSECTS AND CANARIES
medianatures and aesthetics of the invisible

emphasized the decisive role that war plays in technological modernity, and the idea of "total mobilization" as a form of tuning of the national economies, visual culture and personal readiness for war, then what kind of alternative "master narrative" can we find in this slightly different kind of Jünger novel that focuses on microdesign of robotic worlds through animals? This addresses a different kind of readiness, and critique of progress, and still an intertwining of animal energies with technology.

Without going into a fully fledged explication of the narrative – and the protagonist Captain Richard's work application and introduction to Zapparoni's automata factory of entertainment devices – we get a good sense of the slightly different emphasis than in *Storm of Steel*. Through Captain Richard's personal memories and meditations of war – and the coming high-tech war – which are familiar Jünger themes, *The Glass Bees* addresses animal-like automata and the immersive entertainment worlds of such devices that are becoming embedded as part of everyday life. The novel, however, becomes an observation of *obsolescence* – not only in the sense that we think of media device obsolescence in the culture of the "new" media but also replacement of another sort; for the protagonist, this observation becomes clear through his own personal history of war:

> Of course, differences existed between military service under Henry IV, Louis XIII, or Louis XIV, but one always served on horseback. Today the magnificent creatures were doomed. They had disappeared from the fields and streets, from the villages and towns, and for years they had not been seen in combat. Everywhere they had been replaced by automatons. (Jünger 29)

The glass bees are one form of such displacement, replacement and introduction of automation. Described as a mix of a hive and "an automatic telephone exchange," Zapparoni's miniature bee workers represent not only a new form of automatized pseudo-animal labour but a whole system of organization – like a switchboard structure – which automates the carrying of nectar to the hive. As such, as automated independent robots they fulfil Zapparoni's dream of wireless communication networks of semi-autonomous agents (hence, no wonder that Christoph Rosol included a reference to Jünger's novel in his media archaeology of the RFID), but also, for the worried observer, Captain Richard summons the extinction of the organic bee – work and love, in a much-too-perfect balance:

> Bees are not just workers in a honey factory. Ignoring their self-sufficiency for a moment, their work – far beyond its tangible utility – plays an important part in the cosmic plan. As messengers of love, their duty is to pollinate, to fertilize the flowers. But Zapparoni's glass collectives, as far as I could see,

> ruthlessly sucked out the flowers and ravished them. Wherever they crowded out the old colonies, a bad harvest, a failure of crops, and ultimately a desert were bound to follow. After a series of extensive raids, there would no longer be flowers or honey, and the true bees would become extinct in the way of whales and horses. (Jünger 135)

Extinction, replacement, disappearance, innovation coupled to obsolescence are themes that stand out from Jünger's novel, and act also as a trailer to this paper.

Jünger's science fiction world touches on this displacing of the animal as part of the logic of automation and relates to what I have called "insect media": the non-human qualities and models for technology that animal worlds have offered from nineteenth-century entomological discourse to more recent software culture (Parikka, *Insect Media*). In addition, *The Glass Bees* nods towards the double-bind of modernity in terms of technological obsolescence: the paradox of technological society being that it not only produces technology but also gets rid of it at an increasing pace, as well as getting rid of and modulating the organic as part of that process. It also produces obsolescence and nonuse. In such a context of animals and technology, and insect media, we can refer to this as an anti-McLuhan take on media history where technology is not modelled on the human being but has a more complex entanglement with a variety of animal bodies and nature. The approach might differ slightly from the emphasis that Friedrich Kittler suggested in his own anti-McLuhan reminder that before we are able to think media as extensions of man they themselves include a range of other, very non-human processes anyway. Into this mix I want to throw in animals and the idea of how animal bodies are themselves mediatic, and to be approached as aesthetic and material-epistemological figures in order to understand bodies stretched across various differing ecologies.

Indeed, we can claim that there is a wider mobilization of animals and natural resources as part of technological modernity and its forms of perception, related to political economy of media (for instance, electronic waste as one central form of pollution) as well. Also emblematic of the discourse of the posthuman, we are increasingly forced to think of *worlds without the human* – both for ethico-aesthetic and for empirical reasons. Guattari has been the thinker perhaps most contributing to the double notion of ethics combined essentially with new aesthetic paradigms, and the empirical refers here to a possible future that, according to various scientific modelizations, might be true; a world without human life, and various forms of animal life, if our climate change predictions are accurate (see Chun).

Without going into the detail of the various entanglements in which simulations concerning the future of climate change are impacting on the epistemology of the crisis, we can observe a parallel history concerning technology and animals. One way to make sense of this is to take up Akira Mizuta Lippit's argument concerning modernity as intertwining trajectories of animals and technology – where the gradual emergence of technical media during the nineteenth century was paralleled by a specific attitude (and practices) concerning animals. The disappearance of animals from urban cultures of technical media was paralleled by the *appearance* of animals in various discourses, from media (early cinematic discourses being a good example) to modern subjectivity (e.g., psychoanalysis). Disney's mice can be seen as only part of the technological eradication of rodents from urbanity, and the appearance of various animals in scientific films, literature discourses, and animations is part of various measures to control the animal as a production force – and disturbance. The new regimes of media – which were to a large extent used as tools for scientific measurement such as chronophotography and the various measuring instruments of physiology and experimental psychology labs – were ones that tapped into the speed and slowness of animal bodies. Here, we should pay attention to a genealogical understanding of media that does not start from a current bit-too-easy conflation of "media" with entertainment media, but acknowledges that our mediatic devices should be approached

through their archaeologies as scientific measurement devices (see Ernst). The emerging scientific epistemology concerning animal life was feeding into aesthetics in the wider media and popular culture sense; the emerging media technologies such as screen media were instrumental both as producers and mediators of the animal as a specific scientific question.

Hence, this intertwining of animals and technology is not only metaphorical. Instead, the disappearance of the animal is emblematic, measuring impacts and affects of emergence of technicality and, in this case, visibility and invisibility. Indeed, in various contemporary visual and scientific practices, the animal is not only an object of concern but is itself a surface of registration, storage media and a signal of the processes concerning pollution and waste. We literally seem to learn through the "case studies" of animals, whether in news media (as the case of bumblebee disappearance, to which we will return below), scientific data (the discourse of the sixth mass extinction of biodiversity) or other media, including fiction (Coupland) and documentaries (*Vanishing of the Bees*, 2009).

Hence, in order for us to account for this idea of "animal media" as an implicit ethico-aesthetic and epistemological figure, we need to address the entanglement of technical media, animal bodies, and discourses of ecological crisis and waste. Hence the use of the title "insects and canaries" refers to the use of (canary) birds in mining practices as well as the gas warfare of the First World War to detect the presence of dangerous air pollution. Hence, it was indeed not only in mines but also in trench warfare where such ideas were used to detect the impending danger to human lungs as well. In an early test, George A. Burrell of the United States Bureau of Mines conducted tests on various animals, including himself, exposed to carbon monoxide. It took a minute for the canaries to start asphyxiating, eleven minutes for pigeons, twenty minutes for himself, while chickens took no notice (Harrington 259). The canary became metaphoric as a way to transpose invisible, deadly toxins, which in this case I transpose to another layer: to investigate notions of disappearance, obsolescence, the eco-logical crisis and animal organisms in relation to technological epistemologies. In this case, animals such as bees are early-warning systems (with a nod towards the earlier use of the term in media theory by Marshall McLuhan), epis-temological and aesthetic figures of a different kind that work as forms of animal aesthetics. Besides the concept there is another link too: the development of chemicals for gas warfare led to a massive redeployment of such scientific data and resources to pesticide production. This, for its part, has been suspected as one key cause of the bee colony disorders.

Obviously we have had a fair number of theoretical accounts that establish links between technical media culture and animal energies and intensities from cybernetics to Donna Haraway, to such materialist feminists as Rosi Braidotti and Elizabeth Grosz. The various perspectives have paved the way for the so-called wave of "new materialist" thought.[1] A thorough discussion of the various meanings of new materialism remains outside the scope of this text. Instead, I want to flag the usefulness of such projects that aim to think the entwining materiality of temporal bodies – of animal and human – as a question of the ethico-aesthetic. Indeed, of the recent dis-cussions I believe it is Braidotti who has come closest to what I want to argue – that the animal energies, intensities, and productive forces, which are in no way limited to the human, are actually the motor, the energy source, for so much of technological modernity and capitalism. Of course, this energetic per-spective that aims to develop a trans-species ethics, outside an anthropocentric prejudice, is one that also accounts for what it most often means to mobilize animals as part of technologi-cal capitalism. Animals are living matter – and "material for scientific experiments. They are manipulated, mistreated, tortured and geneti-cally recombined in ways that are productive for our bio-technological agriculture, the cos-metics industry, drugs and pharmaceutical industries, and other sectors of the economy" (Braidotti 98). Related to this, we are forced to observe the necessary entropic quality inherent in such a mobilization of animal bodies; that they embody and express a variety of temporal-ities in which their material potentials are being consumed. Indeed, what is discussed in terms of technological obsolescence is a matter of forces of production and consumption more widely too: I am referring here to the exhaustion of energies of living matter, from people (labour) to animals and natural resources. Much of this logic was well summarized in the idea of planned obsolescence introduced in the midst of the Great Depression of the 1920s and 1930s: that products and devices should be legally declared "dead" after a certain period of use, and hence be replaced through legal force. This did not make it onto the statute books but, as we know, it did as part of the mode of production of devices and desires of capitalist consumer culture (Hertz and Parikka). And yet we need to establish the link between technology and where technologi-cal modes of production and consumption draw their resources from, and mobilize as part of the drive for obsolescence, which indeed, as we should realize, is a matter of obso-lescence of animals too.

II

Technology is one part of the wider story con-cerning urbanization and modernization, which play their role in what has been speculated during recent years as the new mass extinction of animal species, including insects. Insects are in this complex ecological pattern – and ecological in the manner that includes various spheres from technology to political economy to nature as well as the ecology of subjectifica-tion in the manner that Félix Guattari argued – in a crucial role because of their centrality for pollination, decomposition and soil proces-sing (Pickrell). The process of "co-extinction" that follows from the loss of one species, piling up as a chain of extinctions, is character-istic of such ecological relations that define a milieu approach to the world: no thing without another, i.e., relations define entities, co-evolution is shadowed by co-extinction, and such processes of co-being and becoming extend much outside the organic. Recent years of Deleuzian-inspired theory of biopolitics and art has picked up on Lynn Margulis's idea of symbiotic co-evolution as well as Bateson's ecology (through Guattari), and now we can extend such ideas of milieu-bound becomings to a grim side of "co-extinction" as well, which also addresses the ontogenetic and material sides of change. Indeed, if relations compose each other (Fuller, *Media Ecologies* 95), they might as well unfold, disperse, and recompose in some other form. This approach also recognizes the longer genealogy of ecological thinking, pre-dating Guattari. For instance, Gregory Bateson's remarks in "Pathologies of Epistemology" are right to the point in their acknowledgement of how the ecological mode of questioning has opened a broader field of consideration of what matters in terms of discourse of nature; from a hierarchical biological focus on family lineages, species, sub-species down to individuals, the ecological as argued by Bateson is where we stop for a moment to consider what exactly is the "unit of selection" (if you want to use Darwinian vocabulary): this makes us think of the couplings of genes with organisms, organisms in environments, ecosystems and if we want to consistently continue, and, as Guattari and the more recent wave of media ecology have done, we need to account for a whole host of "extra-biological" ecologies in order to avoid the epistemological error of "choosing the wrong unit" (Bateson 459) where we start our epistemological inquiry.

Guattari argued in the 1980s that to understand such forms of milieu, or ecology-bound thought, the only real option for the green movement is to extend its concept of nature and animals to include a variety of other spheres. Indeed, as a form of mixed semiotics, he was insisting on methodologies that take into account the variety of ecologies and processes that contribute to processes of subjectification, signification and a-signification. Guattari proposed three ecologies inclusive also of the psyche and the social with their particular "waste" and "pollution." What still makes Guattari's idea refreshing is how it offers a complex material epistemology, completely tied together with aesthetics as a way to think what could be called transversal subjectivities – the shared milieus of articulation for humans and non-humans, and what more recently philosophers such as Braidotti have developed into new forms of (Spinozian) ethics for an age of political economy and technologies of *bios/zoë* – life. To this already complex mix of various ecologies we can add media ecology as one specific field of practices, energies, epistemologies and articulations of the ethico-aesthetic (Fuller, *Media Ecologies*). Through a media ecological focus we are able to investigate how media technological energies contribute to the patterns of replacement, displacement and disappearance of animal energies, and hence hopefully avoid too-narrow "epistemological errors" (Bateson) in our investigation of the aesthetico-epistemologies of disappearance.

Indeed, one of the characteristic features of "animal extinction" is the question of visibility. Besides the obvious point about mediation in popular culture narratives, such doomsday scenarios are, of course, embedded in the larger question of measurement, validation, comparison and presentation of scientific facts so that the radical complexity of such intertwining becomes understandable. This also refers to how we constantly discover new species, which is one part in the contemporary biology of species and populations (Heise, "Lost Dogs"), and emphasizes that despite the fact that it is most probably true that we are in a catastrophic situation concerning animal and natural life, we need more complex ways to understand this situation as relational. Instead of a closed container model of ecology we need to account for natures that are more than objects for measuring visualization, and think of more ontogenetic epistemologies – ecologies as constant creative processes that are entangled with various scientific practices of knowledge production across species and populations. As Ursula Heise in her "Unnatural Ecologies" article reminds us, the conceptualizations of nature and media ecology work both ways, changing perceptions and aesthetics of understanding of nature and technology. Indeed, one crucial question that in a way echoes Lippit's point mentioned above is how the perception of the ecological crisis since the 1960s has been paralleled by increasing media theoretical talk of media ecology (for instance Neil Postman), as if hinting that the disappearance of the natural ecology has its counterpart in technological conceptualizations. In any case, we are safe to say that the two are very much intertwined, and perhaps never were detached in the first place. In her usual perceptive manner, Wendy Hui Kyong Chun argues that the whole epistemology concerning scientific simulation is itself a question of how we relate to data, software and programming which involves a curious relation to the future: climate predictions are not untrue because they are extrapolated from a massive amount of data that speculates a possible future, and allows us a possibility to act on that one particular data epistemology – or even aesthetico-epistemology, as it involves various forms of visualization and aesthetics too as part of its software-embedded knowledge production.[2]

But mass extinction is not just something directly observable, and includes the difficulty of detection (Heise, "Lost Dogs"; see also Pickrell). The massive scale of climate change that involves attempts to offer a convincing epistemology by tying up pasts (data collected over decades) with futures (patterning data to offer a premediated scenario) is paralleled by the problems of detection that biologists have to face. This relates to the status of sub disciplines in biology. Of these, molecular genetics enjoys a prominent role in defining what a species is, and the re-emergence of taxonomy, as Heise ("Lost Dogs" 55) notes. In short, we are faced with questions not just of going out there and observing a situation but also having first to address how we in the first case talk about animals, species, populations and extinction. This question ranges from disciplinary knowledge and practices to a wider set of narratives, technology and interests of knowledge (ibid.): a technologically embedded material episteme that is itself entangled (in Karen Barad's way of using the term to avoid a Kantian correlationism) with the objects of knowledge it produces. In a situation in which we still have not even

documented many insect and other species, we are faced with the challenge of a possible mass extinction underway, but which is difficult to turn into an epistemology with, let's say, policy impact. When dealing with populations and species, biologists are faced again with similar problems of data collection from empirical and historical sources, and translating that into modes of perception that are convincing from an aesthetic and epistemological point of view. Again, to emphasize, aesthetics refers here not to ornamentality – or even science communication – when talking of visual communication of scientific facts, but to a more fundamental role that perception plays in all this.

A widely media-reported environmental issue of recent years has been the mysterious bumblebee extinctions especially, in the United States but also worldwide – reported probably because not only of its catastrophic implications but the cuteness of the subject topic. In the same category of cuteness as pandas, puppies, penguins and dolphins, and hence suitable for gentle discourses concerning preservation, bees have been addressed as one of the most recent victims of climate change. What started as a mysterious wave of mass deaths of anything up to 60–70 per cent of the bee population in certain parts of the United States spread to become a worldwide phenomenon, the cause of which remains a mystery. Whereas estimates were of an apocalyptic scale for the bees themselves – that at this pace bumblebee species would be wiped out in a few years' time ("Bumblebees Could Face Extinction") – this was registered as apocalyptic on another level too: no bees, no pollination; no pollination, no crops; no crops, no human beings.

Despite media hype tailing off in recent years, in 2010 the United Nations Environment Programme released an alarming report entitled *Global Honey Bee Colony Disorders and Other Threats to Insect Pollinators* that mapped the grim picture:

> Current evidence demonstrates that a sixth major extinction of biological diversity event is underway. The Earth is losing

between one and ten percent of biodiversity per decade, mostly due to habitat loss, pest invasion, pollution, over-harvesting and disease. (1)

The data collated in the report not only registered a sudden slump in the honey-producing colony numbers but also, and perhaps more worryingly, a steady decline over a longer period. Indeed, what such diagrammatizations have to deal with, in terms of the aesthetico-epistemological modes of perception, is to arrange time-scales; the past ten years of public discourse have produced and fed on narratives of the catastrophe of the sudden decline in bee populations, but the narrativization of a longer decline is still something that is not so easily or willingly picked up. Indeed, as Heise notes of various earlier forms of media ecological theoretization, the notion of environment used in media theory implied "a spatial perception or experience" ("Unnatural Ecologies" 165). And yet, with the contemporary discourses and epistemological practices of climates, extinction and relations between animals and technology, we are increasingly faced with the question of how to think/visualize/narrativize time in terms of non-human scales. Indeed, to quote Heise:

> Questions of scale also matter for the stories we tell about biodiversity in other ways. Human perception and cultural understanding of species loss normally focus on the orders of magnitude closest to us, whereas processes at other scales often do not make their way into the public consciousness. ("Lost Dogs 57)

This is where anthropocentric perspectives fail to grasp the mixed ecological milieu, across species, and humans and non-humans – and a more transversal ethics of perception is needed (Braidotti). One hundred crop species provide 90 per cent of human food worldwide. Of these one hundred, seventy-one are bee pollinated, showing the curious way in which our survival is very much tied together with the bees. Such narratives are of great use in rescaling issues of massive spatial and temporal scales to make sense, and hence create certain aesthetics of such a crisis too.

Such crises participate in a longer cultural history of narrativization of nature vs. humans/modernization (Heise, "Lost Dogs"), but what I want to focus on is the entanglement with technical media, and, as promised, the relation to aesthetics. In other words, a trans-species ethics also needs to be a *transmaterial ethics*, which takes into account technologies, material epistemologies, scientific practices, aesthetic discourses concerning the entanglement of technical media and animals, etc. This does not always refer only to the established list of what we count as "media" but also such scientific technologies as DNA fingerprinting too; in this curious case of the bees, such techniques were used to map information concerning colonies and their relations, and the suggested causes most often took into account pollution and other modernization-related effects, showing the further entanglement of the bee question in the wider technological modernity. This is where I want to nod towards the title of the text again, namely "insects and canaries." The use of pesticides evinces the co-evolving history of war and control of nature, as Russell (2) argues: the entanglement of techno-scientific development as well as organizational/industry arrangements between military and business, supporting the mobilization of early gas warfare into insect pesticides, and back to human warfare. Identification of the use of certain pesticides as contributing to the recent bee colony disorders is one of the more convincing causes, and also presents this curious link back to discourses of ecocrisis – across histories of war, animals, and techno-scientific developments.

As to the various other causes suggested, one often mentioned but that still lacks data has to do with electromagnetic radiation and the technical communication sphere of such devices. It's not just that bees are part of media history, but also that they are themselves mediatic. Their body incorporates a crystal that contains lead, and hence is receptive to electromagnetic communications – the regime of communications that works through Hertzian vibrations. High-frequency mobile communications, including, for instance, RFID (Radio Frequency

Identification Devices) have a level of resonance with the communication of the bees – the famous bee dance, discussed, for instance, by Karl von Frisch in the middle of the twentieth century (see Parikka, *Insect Media* 121–44) – that takes place at frequencies of 200 Hz and 300 Hz; GSM (Global System for Mobile Communications), for example, has a very different carrier frequency (800–2200 MHz) but the pulse frequency fits exactly into the slot – 217 Hz.

I am less interested here in the question of whether or not this link is the true cause; whether the little world of communication, the anti-McLuhan global village of insects, is really being disturbed by the parasitical human wireless communications. Instead, I am fascinated by the sheer fact that this connection is being suggested – not only in this context, of course, but also in the various studies that link that pulse frequency to disturbances in the brain waves of people. The fact that people are making these links and exactly through a theme of pollution (of the electromagnetic spectrum) is of significance in terms of understanding continuums between nature and culture, or, as I have called it recently, *medianatures* (Parikka, *Medianatures*); the inherent link that mediatic regimes and high-tech cultures, by necessity, have to nature, the animals, and materiality of such regimes, also through waste and pollution. It is one way to investigate the notion of media ecology, as well as the mediatization/aesthetics of "natural" ecology. Indeed, this demonstrates the further infiltration of technology in the epistemology and discourse concerning animals and mass extinction, part of the wider debates and research concerning our large-scale ecocrisis. Such links elaborate on the need to develop new methodologies to track the continuum between animals and humans, technology and ecology, political economy to technology, aesthetics to ecocrisis.

Indeed, the curious question as to the scale and causes of a variety of disturbances in animal and "natural" life is constantly embedded in the frameworks of knowledge needed to elaborate the *not immediately perceptible*. What has been a constant theme in modern technological aesthetics – the intertwining of animal forms of perception and technological forms of perception both as non-human worlds – is now also something that can help us think the various regimes of knowledge concerning the current ecocrisis, disappearance and obsolescence. Indeed, again in the manner that Chun argues in her "Crisis, Crisis, Crisis" article concerning the epistemology of climate change and technological simulations, and in the manner of ongoing modern curiosity in terms of the microworlds (or just alternative worlds) of animals (why they communicate, sense, perceive so differently), we are confronted with the need to think through the animal and the non-human. In the context of the ecocrisis and scientific knowledge, we are similarly engaged in this double bind of animal worlds of perception, and the radically different aesthetics of animals, as well as the aesthetics and modes of perception, afforded by complex technological forms – for instance the Geophysical Fluid Dynamics Laboratory (Chun 107).

In this sense, Jünger's "glass bees" are quite an apt literary example that addresses the disappearance of bees; the double bind of technological modernity as part of animal worlds is not, however, only a theme of obsolescence understood through the military metaphors of an arms race (even chemical), and the changing face of the technological-scientific army, but the wider media sphere. This point becomes evident when analysing the material constitution of our screen technologies and their e-waste load, as Sean Cubitt has been doing, as well as through an analysis of the aesthetics of the animal – both about, but also stemming from, the animal. The media and the natural ecologies are also entangled on another level besides that of the metaphor and narrative.

III

As a way of investigating the link between media, animals, bees, extinction and aesthetics I want to turn to artistic methodologies. Indeed, an increasing amount of artistic work has taken up the posthuman question. There has been a wide range of responses to the "question of the animal" (Wolfe) in the contemporary art sphere, picking up Jacques Derrida's writings concerning the self and the non-human animal (the famous example of the cat and Derrida's nude body). I want to use this aesthetic perspective to elaborate one further angle to the aesthetico-epistemology of the knowledge and disappearance of animals. In this context, I am addressing Lenore Malen's work and especially her 2009–10 piece *The Animal That I Am*. Through a three-screen video installation, it articulates various aspects of the Colony Collapse Disorder discussed briefly above. In Malen's video installation this is the case from the point of view of beekeepers, but also raising various aesthetico-ethical themes concerning the relationship between bees and human cultures. As such, it is emblematic of the technical media and artistic media responses to such developments as the mass extinction of species, and articulates, in its own way, the double bind of technology – including screen media – and disappearance of the animal.

Actually more interesting than the narrative of beekeepers about the universal harmony of the insect world – similar tropes were used at various times in history, even during Germany's 1930s National Socialism when Maya the Bee was the ideal Nazi supporter due to her loyalty

– are the sounds, rhythms and vibrations that Malen introduces as audiovisual elements. The installation surrounds you, through its enveloping soundscapes and its compound images. As such an audiovisual ecology *The Animal That I Am* is an implicit suggestion towards a slight variation: the *Animal* Media *That I Am*. Modulations of perception through media technologies share much with the animal worlds, and the event of cohabitation that the piece tries to offer is one where we are invited to hear, sense, and tap to the rhythmic vibrations – the Hertzian world – of insects. As such, it opens to a slightly more radical non-human stance when you close your eyes and embed yourself in a rhythmic sonic ecology/epistemology.

Also, the three screens (see Fig. 1) of *The Animal That I Am* are rhythmic elements that deterritorialize our vision. A slowly progressing multiplication of viewpoints is the becoming-animal of perception that the installation aims to deliver. The immersive space is also one of composed fragmentation into the compound vision of insects. Slow disorientation is one tactic of this mode of becoming; it points both to the world of insects and to the media in which we are immersed. The early avant-garde connection between the technical vision machine and the insect compound machine – in the words of Jean Epstein, "the thousand

Fig. 1.   From Lenore Malen's video installation. Image reproduced by kind permission of the artist.

Fig. 2.   From Lenore Malen's video installation. Image reproduced by kind permission of the artist.

faceted eyes of the insects" (115) – creates a sense of space as split; perspective is multiplied into a variation. Malen's *The Animal That I Am* is about such forms of multiplicity, but transporting themes familiar from early twentieth-century aesthetics into the contemporary context of bee disappearance. Hence, one cannot avoid asking what is the double bind implied in the installation; the theme of disappearance addressed through the video visions.

Lenore Malen's *The Animal That I Am* intertwines the various histories, aesthetics, and idealizations of the bee community as well as the bee's relations with beekeepers. Donna Haraway's term for this – companion species – comes to mind, but not without friction when you ask how one establishes relations with such insect forms of life as bees. As flagged above, our relation to insects is reflected in much more than the narrative aspect of Malen's work. The immersive environment of the installation envelops the spectator in triggering ways. The clips that Malen uses are mini-thoughts, mini-brains, which are brought together with her digital software tools; the clips are memes that Malen excavates from online archives and audiovisual repositories, and composes into a three-channel envelope. *The Animal That I Am* poses the question:

can insects be our companion species? This is paradoxical in light of Derrida's *The Animal That Therefore I Am*, to which Malen's title refers. Derrida starts with the gaze of the animal – his cat, to be exact, lazily gazing at Derrida's naked body. But catching the insect's compound eyes is more difficult. For Malen, Derrida's essay functions as a critique of subjectivity but we need to account for further levels on which the question of aesthetics and perception features in our relation to animals and ecologies. Indeed, such key posthuman debates in philosophy have been addressing the co-constituting nature of watching/ being watched as a cross-species mode of subjectification for the human and its relation to the non-human. As a further question, we need to ask: what forms of aesthetics and "watching" do we need to carve out in order to understand the other scales of ecology in which we are embedded, being co-constituted not only by cats and dogs but also by complex ecologies in which we co-live, and might also co-extinguish? Such speculative, philosophical and aesthetic questions might give insights into a more complex epistemology of technological aesthetics too. This is exactly why we need to account for the wider framework in which the ecological is given to us, as technical media, and ask how this link of technological, material epistemology is guiding a specific way of thinking insects too as media. The disappearing insect becomes a form of transmission as well as signal of wider ecological connections and chains of disappearances, in a manner in which Douglas Coupland continued this line of thought in his fiction novel *Generation A*: after the near future disappearance of bees, five people in different parts of the world got stung by a bee, and hence are themselves suddenly transmitters, signals, or at least some kind of condensation point for a whole range of measures of scientific concern; what is it in these spatially dispersed people's organisms that attracted the bee? What's more, it points as if to a whole substrate of communication between animal and human bodies, around which a whole scientific and popular cultural (the bee-stung people become media stars) world is summoned:

> I began to imagine the lives of those bees that survived over the years just long enough to find us and sting us and send us their message, to tell us their story. I began to imagine small cells of them – not even hives – surviving from year to year, nesting under highway overpasses and the dusty eaves of failed shopping malls – foraging for pollen in the weeds growing alongside highways, their wings freezing and falling off in the winter and in the summers their wings rotting and leaving them crippled as they tried to keep their queens alive, finding little comfort in each other, finding solace only in the idea that their mission might one day succeed, that they would one day find us, with our strange blood [...]. (Coupland 297)

IV

To conclude, let's return to the original idea about "canaries," or how the question of the animal is itself a measure of our situation concerning technological modernity; from media technologies (including electronic waste) to urbanization, modern agriculture, pollution, and so forth. This conceptualization is tied to a much more complex ecology of things and relations than just an index of how good nature is destroyed by bad culture. The question concerning aesthetics and contemporary art – even visual arts – becomes more interesting when you step out of the representational sphere to tap into measurements and mediations of other sorts. More than art about animals, perhaps we should pay attention to art by and for animals – to use Matthew Fuller's ideas – where the non-human animal question is taken as an aesthetic cue (Fuller, "Art for Animals"). Fuller identifies a two-fold danger in relation to art with/about nature: that we succumb to a social constructionism or that we embrace biological positivism. And yet we need to be able to carve out the art/aesthetic in and through nature and animals in ways that involve the double movement back and forth between animality and humanity. Art for animals is one way to achieve that productive dynamics, to quote Fuller (269): "Art for animals intends to address the ecology of capacities for perceptions, sensation, thought and reflexivity of animals." What's more, this aesthetico-epistemological task is connected to wider possibilities, that

> make us imagine a nature in which nature itself must be imagined, sensed and thought through. At a time when human practices are rendering the earth definitively *unheimlich* for an increasing number of species, abandoning the human as the sole user or producer of art is one perverse step towards doing so. (Ibid.)

What such a perspective raises is not a focus merely on animals but the non-human energies and potentials of/in aesthetics, including media technological aesthetics.

This experimental connection between aesthetics and imagining natures picks up on non-representational notions of art and animality that, for instance, Elizabeth Grosz (*Chaos, Territory, Art*) has emphasized more recently. Nature and animals already are aesthetic, vibrational, erotico-aesthetic milieus of rhythms; and where aesthetics happens, much beyond the human eye gets involved. In artistic practices, David Dunn's bioacoustics can be seen elaborating on similar issues, and scaling the question of animals, nature and aesthetics to that very non-human level too.

The signalling worlds of bees dancing, insect worlds of acoustics, rhythm, and vibrations are in themselves already part of the world of eco-aesthetics, which as a regime is primary to any kind of mediations concerning the displacement (or extinction) of the animal. Indeed, I suggest that such themes should not be taken to strengthen polarities of innocent nature raped by bad technological modernity, as Ursula Heise ("Lost Dogs") argues so much of environmentalist narrativization of extinction has done during the past two hundred years, but should help us to develop new modes of understanding the media–nature continuum as medianatures. This concept is one suggestion to think of the ecological entwinings of epistemology and aesthetics in the context of the ecocrisis, and even, perhaps, one way to address invisibility, disappearance and obsolescence. I want to argue that *disappearance* does not merely flag a theme of "extinction" but also such modalities that we need to struggle to perceive – worlds of non-human perception. This is not to downplay scientific research concerning the de facto disappearance of animals from the world, but to focus on such ecological intertwinings where aesthetics – making things visible – is something that needs to be addressed on a non-human level too. In other words, this scientific level is also dealing with the difficulties of perception, of aesthetics, of addressing so many scales of interaction. Such a multiscalar mapping would necessarily be an ecological project in the manner Guattari proposed; transversal entanglement of technological epistemologies and practices, aesthetic modes of knowledge, non-human ontologies, and awareness of political economy and exhaustive global capitalist production and consumption. In a manner similar to the way in which our mediatic culture is increasingly defined by non-visibles that range from electromagnetic transmissions to algorithmic image processing as the non-visible generation of what we see, we need to extend this to ecology and natures too, embedded in aesthetic "practices." Such art that is able to tap into the intertwining of invisibilities and the unrepresentable complexity is the most interesting and up to date in trying to understand why and how ecocrisis is an aesthetic crisis.

notes

1 Despite the attachment to the discourse of new materialism, Grosz, for instance, has wanted to specify her approach to prefer

> to understand life and matter in terms of their temporal and durational entwinements. Matter and life become, and become undone. They transform and are transformed. This is less a new kind of materialism than it is a new understanding of the forces, both material and immaterial, that direct us to the future. (*Becoming Undone* 5)

2 On climate change, data and simulations, see also Edwards.

bibliography

Barad, Karen. *Meeting the Universe Halfway: Quantum Physics and the Entanglement of Matter and Meaning.* Durham, NC and London: Duke UP, 2007. Print.

Bateson, Gregory. *Steps to an Ecology of Mind.* St Albans: Paladin Frogmore, 1973. Print.

Braidotti, Rosi. *Transpositions: On Nomadic Ethics.* Cambridge: Polity, 2006. Print.

"Bumblebees Could Face Extinction." *BBC News* 5 May 2001. Web. 7 Feb. 2013. <http://news.bbc.co.uk/1/hi/sci/tech/1314012.stm>.

Chun, Wendy Hui Kyong. "Crisis, Crisis, Crisis, or Sovereignty and Networks." *Theory, Culture and Society* 28.6 (2011): 91–112. Print.

Coupland, Douglas. *Generation A.* London: Heinemann, 2009. Print.

Cubitt, Sean. "Ubiquitous Media, Rare Earths – The Environmental Footprint of Digital Media and What to Do About It." Pervasive Media Lab, University of the West of England. 22 Sept. 2009. Unpublished Talk.

Edwards, Paul N. *A Vast Machine: Computer Models, Climate Data, and the Politics of Global Warming.* Cambridge, MA: MIT P, 2010. Print.

Epstein, Jean. "Fernand Léger." *Écrits sur le cinéma tome 1 (1921–1947).* Paris: Seghers, 1974. Print.

Ernst, Wolfgang. *Digital Memory and the Archive.* Ed. and intro. Jussi Parikka. Minneapolis: U of Minnesota P, 2013. Print.

Fuller, Matthew. "Art for Animals." *Deleuze/ Guattari and Ecology.* Ed. Bernd Herzogenrath. Basingstoke: Palgrave, 2009. 266–86. Print.

Fuller, Matthew. *Media Ecologies: Materialist Energies in Art and Technoculture.* Cambridge, MA and London: MIT P, 2005. Print.

Grosz, Elizabeth. *Becoming Undone: Darwinian Reflections on Life, Politics and Art.* Durham, NC: Duke UP, 2011. Print.

Grosz, Elizabeth. *Chaos, Territory, Art: Deleuze and the Framing of the Earth.* New York: Columbia UP, 2008. Print.

Guattari, Félix. *The Three Ecologies.* Trans. Ian Pindar and Paul Sutton. London and New Brunswick, NJ: Athlone, 2000. Print.

Haraway, Donna. *When Species Meet.* Minneapolis: U of Minnesota P, 2008. Print.

Harrington, John Walker. "The Canary Birds of War." *Popular Science* 93.2 (Aug. 1918): 258–60. Print.

Heise, Ursula K. "Lost Dogs, Last Birds, and Listed Species: Cultures of Extinction." *Configurations* 18.1/2 (2010): 49–72. Print.

Heise, Ursula K. "Unnatural Ecologies: The Metaphor of the Environment in Media Theory." *Configurations* 10.1 (2002): 149–68. Print.

Hertz, Garnet, and Jussi Parikka. "Zombie Media: Circuit Bending Media Archaeology into an Art Method." *Leonardo* 45.5 (2012): 424–30. Print.

Jünger, Ernst. *The Glass Bees.* Trans. Louise Bogan and Elizabeth Mayer. New York: New York Review of Books, 2000. Print.

Kittler, Friedrich A. *Gramophone, Film, Typewriter.* Trans. Geoffrey Winthrop-Young and Michael Wutz. Stanford: Stanford UP, 1999. Print.

Parikka, Jussi. *Insect Media: An Archaeology of Animals and Technology.* Minnesota and London: U of Minnesota P, 2010. Print.

Parikka, Jussi, ed. *Medianatures: The Materiality of Information Technology and Electronic Waste.* Open Humanities Press, 2012. Web. 7 Feb. 2013. <http://livingbooksaboutlife.org/>.

Parisi, Luciana. "Technoecologies of Sensation." *Deleuze/Guattari and Ecology.* Ed. Bernd Herzogenrath. Basingstoke: Palgrave, 2009. 182–99. Print.

Pickrell, John. "Mass Extinction of Insects May be Occurring Undetected." *National Geographic News* 20 Sept. 2005. Web. 7 Feb. 2013. <http://news.nationalgeographic.com/news/2005/09/0920_050920_extinct_insects.html>.

Rosol, Christoph. *RFID. Vom Ursprung einer (all) gegenwärtigen Kulturtechnologie.* Berlin: Kadmos, 2007. Print.

Russell, Edmund. *War and Nature: Fighting Humans and Insects with Chemicals from World War I to Silent Spring.* Cambridge: Cambridge UP, 2001. Print.

Sterling, Bruce. "Introduction." *The Glass Bees.* By Ernst Jünger. Trans. Louise Bogan and Elizabeth Mayer. New York: New York Review of Books, 2000. vii–xii. Print.

United Nations Environment Programme (UNEP). *Global Honey Bee Colony Disorders and Other Threats to Insect Pollinators.* New York: UNEP, 2010. Print.

Wolfe, Cary. *What is Posthumanism?* Minneapolis and London: U of Minnesota P, 2010. Print.

exhibit 1: the green-eyed monster

Envy and jealousy are the shameful private
parts of the human soul.

Nietzsche, Human, All Too Human

Man never understands how anthropomorphic he is.

Goethe, Maxims and Reflections

Without question, jealousy looms over
the reader as one of the greatest and
most enduring of literary thematic shadows.
The "green-eyed monster" has stalked the
verses of Homer and the pages of Shakespeare
up to the hyper-neurotic chroniclers of today.[1]
Jealousy itself, however, was both exacerbated
and complicated in nineteenth-century Europe,
thanks to rapidly changing social and economic
circumstances – the "woman question" being
one of the most salient.[2] Tolstoy's remarkably
economical novella *The Kreutzer Sonata*
manages to create one of the most intense,
vivid, and thought-provoking portraits of jealousy in the canon, and is as disturbing to read
today as it no doubt was in 1889 (especially if
you happened to be Tolstoy's wife, Sofia
Andreyevna, who – despite, or *because* of,
appearances – promptly lobbied the Tsar
himself, to have the official ban on this work
overturned, so that people would not presume
the portrait was based on her marriage to the
author). The madman's tale – and we will not
waste our breath speculating on degrees of
sanity or insanity – is at once compelling,
counter-intuitive, carefully reasoned, and contradictory. Without completely identifying
Pozdnyshev as the author's ventriloquist
dummy, there are certainly many moral and
ideological parallels between the ideas of the

dominic pettman

TOLSTOY'S BESTIARY
animality and animosity in the kreutzer sonata

fictional murderer and the real patriarch;
many of these explicitly flagged by Tolstoy
himself in the subsequent epilogue to the
story, that he felt obliged to write in order to
underline the moral of the story (for those
rather slow-witted souls who could not see it
written in blood on the surface of the text
itself, before even seeping down into the
subtext). The lesson to be learned revolves
around the value of "continence," not only
regarding women in general but even toward
one's own wife. Celibacy is presented as an
ideal to which all civilized human beings
should aspire, in order to more faithfully
follow Christ's teachings; for carnal appetites –
even the lawful kind, sanctioned by

marriage – lead to sensuality, which leads to infidelity, which leads to jealousy, which potentially leads to irreparable violence. As Pozdnyshev himself puts it: "Of all the passions, it is sexual, carnal love that is the strongest, the most malignant and the most unyielding" (48). And so we meet the first of thirteen exhibits in Tolstoy's bestiary: the green-eyed monster. This unpleasant affect, this "rabid beast of jealousy" (115), is a creature we shall merely introduce at this point, and examine in more detail later in the tour, after meeting some more of such zoological specimens, themselves embedded in a book which is figuratively bursting at the seams with animal metaphors.

While sitting in a train carriage, traveling through the Russian night, Pozdnyshev tells his terrible tale to the narrator in an almost uninterrupted monologue. The homicidal climax approaches as inexorably as the destination of the journey, and as inevitably as the dawn which will greet the weary travelers.[3] The first movement of the novella, however, is comprised of a discussion between decent citizens riding the same train – "a plain, elderly lady," a lawyer, and an old man. This conversation concerns the rapid mutation of social circumstances, especially the new status and visibility of divorce. The older gentleman blames the education of women for all the trouble and fuss surrounding modern "marital discord." The others are both amused and appalled by this man's reactionary reasoning,[4] the woman responding, "After all, it's only animals that can be mated at their master's will; human beings have inclinations and attachments of their own" (8). "You're wrong there, missus," replies the old man. "The true difference is that an animal's just an animal, but human beings have been given law to live by."[5] The young woman counters, "but I think you would agree that a woman is a human being, and that she has feelings just as a man has, wouldn't you? So what's she supposed to do if she doesn't love her husband?" (10). The old man fends off her objections by referring back to the animal kingdom: "If he'd never given her any leeway in the first place but had kept her properly reined in, she'd no

doubt still be living with him to this day. You mustn't allow them any freedom from the word go. Never trust a horse in the paddock or a wife in the home" (11).[6] Indeed, this observation foreshadows a key moment later in the story, when Pozdnyshev becomes most concerned about his wife at the very moment she has learned the malefic art of contraception, and thus morphs into a threatening equine state in his own mind: "She was like an impatient, well-fed horse that has had its bridle taken off, the same as ninety-nine percent of our women. I could sense this, and it scared me" (83).

It is this ambiguous tension which will serve as the focus of this present essay, that is, the tug-of-war between the animal and the angel, in which the human being replaces the role of the rope (the human being, in this instance, itself being little more than a tug-of-war between the genders).[7] Scarcely a page of *The Kreutzer Sonata* neglects to set up and then develop an explicitly coded, and morally loaded, distinction between humans and beasts. The logic behind this common form of ontological apartheid, however, becomes less and less coherent the closer we seek to examine it. And the motivation for doing so is not to fault Tolstoy for *his* fuzzy logic alone – since even his contemporaries found the author's theo-philosophical writings wanting in comparison to his literary efforts – but to show how this kind of species-based policing continues into the present, despite pressures from all sides to the previously unassailable confidence of human exceptionalism. In other words, this essay deals with the ways in which implied definitions of "animality" and "animosity" represent two sides of the same coin in Tolstoy's libidinal economy; and, moreover, how these two terms are deployed in the text to clear a pure zone for the potential salvation of human souls, at some future time, at the ultimate expense of nonhumans. As the meta-narrator crucially observes, "this *animosity* was nothing other than the protest of our human nature against the *animality* [*ozloblenie*] that was suffocating it" (56; my emphasis).

For Pozdnyshev, as for Tolstoy, "man" exists under the ethical imperative to strive for a truce in the battle of the sexes, whereby men respect women enough not to lust after them, ideally not to impregnate them, and certainly not to continue venal relations in the unfortunate case of conception. ("Man" here includes woman, since Tolstoy sees a perverse equality in the susceptibility of both genders to "becoming-animal," figured – in contrast to Deleuze and Guattari – in a negative sense.)

Such are the debauched behaviors of the average citizen, magnified in depravity as one climbs the social ladder, the worst offenders being the upper classes, thanks to the obscene hypocrisy involved, and their covert conscription of the poor into their own therapeutic pleasures.[8]

exhibits 2, 3 and 4: the fornicator, the reptile-doctor, and the venus fly-trap

And so we meet the second exhibit in Tolstoy's bestiary: the fornicator. The fornicator is the monstrous product of an emerging biopolitical regime, in which the "solicitous government" is in cahoots with the medical establishment, the class structure, as well as the patriarchal institution of the family, in order to encourage and regulate the libidos of young men in the interest of the homeostatic health of the body politic. Through a combination of gluttony and sloth, the bourgeoisie are the all-too-willing prey for "a systematic arousal of lust" (34), although, truth be told, even the *muzhaks* have "fallen" without the same decadent stimuli. In Pozdnyshev's account, the erotic insights of Freud and Reich are anticipated by the Russian state (as well as in other countries throughout Europe), and pre-emptively controlled in the form of "tidy, legalized debauchery" (24). This in turn creates a new type of male subjectivity, as Foucault could also have well told you (although without the moral disapproval): a kind of artificial hybrid between goat and peacock. Thanks to the "reptile" witch-doctors – our third member of Tolstoy's bestiary – and their obsession with public hygiene via private parts, "simple,

natural" relations with women have been ruined, perhaps forever (25). As the protagonist tells us, "A fornicator may restrain himself, struggle for self-control, but never again will his relation to women be simple, clear, pure, that of a brother and sister" (25).

Which leads us already to the fourth exhibit in our monstrous menagerie (since they are all related, actively encouraging the others into very existence): the Venus fly-trap. Pozdnyshev complains:

> Women know perfectly well that the most elevated love – the most "poetic," as we call it – depends not on moral qualities but on physical proximity and also on things like hairstyle, or the color and the cut of a dress [...] She knows that our man's lying when he goes on about lofty emotions – all he wants is her body, and so he will willingly forgive her the most outrageous behavior. What he won't forgive, however, is an outfit that is ugly, tasteless or lacking in style. A coquette's knowledge of this is a conscious one; but every innocent young girl knows it unconsciously, as animals do. (31–32)

Tolstoy, the Puritanical Christian, even inserts an ironic reference to women who use the writings of Darwin as bait for our presumably progressive suitor-cum-fornicator. "Ah, the origin of species," trills the young nubile, "how interesting!" Pozdnyshev is thus quite blunt in his insistence that "Marriages nowadays are set like traps" (36). What we thus see is the author's diegetic recognition of a generalized "technopoetics of capture" deployed by women in order to snag a socially acceptable, if not beneficial, mate; their clothes, their cosmetics, their somatic grammar, all simultaneously manifesting the cause and effect of "the prism of our artificial way of life" (35). Indeed, the *technical* aspect of this massive capture operation is made explicit in Pozdnyshev's descriptions of women instrumentalizing their own social alienation and sexual objectification into a perverse, reverse form of allure. He stresses that "it's this domination by women we're suffering from, it all stems from that."

> "What domination?" I asked. "All the rights and privileges are on the side of men."

"Yes, yes, that's just the point [...] Women are exactly like the Jews, who by their financial power compensate for the oppression to which they're subjected. 'Aha, you just want us to be merchants do you? All right, then, it's as merchants that we'll lord it over you,' say the Jews. 'Aha, you just want us to be objects of your sensuality, do you? All right, then, it's as objects of your sensuality that we'll enslave you,' say women." (39)

The pardoned murderer describes this particular master–slave dialectic in terms of a "technique" which when abused "acquires a terrible power over men" (40).[9] Moreover, "Women have turned themselves into such effective *instruments* for acting on our senses that we can't even speak to them with equanimity" (41; my emphasis).[10]

These biotechnologically enhanced Venus flytraps are at once predator and victim, as we see Pozdnyshev's (and by extension Tolstoy's) circular logic in full force. On the one hand they are the height of civilized artistry, and on the other they are behaving according to unconscious animal instincts. (Remembering that "human nature" is the ultimate oxymoron.) For women to embrace their full humanity, they must not seduce, consciously or unconsciously. They must not play the game, for sex itself is an inescapable vice for our species: whether you figure us as the children of Adam, or the children of Darwin. And the irony of this is not lost on our narrator, trying to follow the curvature of this slippery line of argumentation.

"Why vice?" I asked. "I mean to say, you're talking about one of the most natural human activities there is!"

"Natural?" he said. "Natural? No, I tell you, quite the contrary's true, I've come to the conclusion that it isn't ... natural. (46)

And what is Pozdnyshev's evidence? The fact that young women are traumatized by it, especially on their wedding night, if brought up in a good home. The madman continues, "Eating is natural. Eating is something joyful, easy and pleasant which by its very essence involves no shame. But this is something loathsome, ignominious, painful."

The confusion is highly symptomatic: animals are figured as the lowest of the low in one gesture – subhuman, in other words. Yet they are also held aloft as the avatars of Nature, on the other, from which humanity has become exiled or estranged. The animal pulls double-duty as disgusting example and noble ideal. Like "the primitive," of which this period was also enthralled, animals are at the same time guilty (of not being civilized enough) and innocent (of the crimes committed by so-called civilization). Now this would make sense if contemporary Europeans divided themselves ideologically on one side or the other, but the fact is that in the majority of thinkers – and especially vividly in Tolstoy – this ambivalence is problematically co-present in the same mind, and in the same argument. (Witness, for instance, the absurdly confused and earnest notion that privileged children have become unnaturally sensual because they have been raised "like the young of animals." That is to say, "The attire, the reading, the shows, the music, the dances, the sweet food, the whole circumstance of life, from the pictures on the boxes to the novels, stories, and poems" – as if bunnies or piglets are exposed to such things!)[11] And yet there remains *some* kind of graspable parallel at work, in which the overfed animal becomes a freak of nature, just as the over-stimulated child becomes less-than-human. Such an equation would be at least internally consistent if Pozdnyshev and/or Tolstoy could resist switching inconsistently between their figuration of nonhuman rhetorical devices.

For the author himself, as expressed in the belated epilogue, carnal love is an "animal condition." ("Yes, I was a dirty pig, and I thought I was an angel" (42).) And yet, as we have seen, it is enabled and encouraged through the most exquisite forms of *human* technology. Nothing less than humanity's much-lauded, but rarely seen, "human dignity" is at stake (a condition which is, according to Tolstoy, an asymptotic ideal, but no less important to strive toward, for all that).[12] Unlike pigs and rabbits, humans do not exist merely to "reproduce themselves as abundantly as possible," nor should they succumb to the temptations of "monkeys and Parisians," who shamelessly "enjoy sexual pleasure with the greatest degree of refinement possible" (48).[13] The only answer is to refrain from sexual intercourse altogether.[14]

exhibit 5: the porcine couple

And at this fork in the road we meet the next and fifth exhibit in Tolstoy's toxic zoological gardens: the porcine couple. Pozdnyshev is a great cynic when it comes to love: he denies that this all-important affect is anything more than a poetic form of camouflage for the most debased of animal instincts. Soon after his marriage to the woman who remains nameless, the reality of the situation becomes disturbingly clear to him: an experience he extends to all his fellow men and women, assuming they do not see this reality so clearly, due to the spell of denial which the troubadours of love have woven throughout society. Pozdnyshev could thus be caricaturing Kant's infamous statement that the wedding ceremony functions primarily as a way to ensure that the happy couple enjoy exclusive and mutual *legal* access to each other's genitals. "Our amorous feelings for each other," he recalls painfully, "had been drained by the satisfaction of our senses, and we were now left facing each other in our true relation, as two egotists who had nothing whatever in common except our desire to use each other in order to obtain the maximum amount of pleasure" (52). A secular sin, of course, in Kant's schema, since humans – as opposed to animals – are never to be utilized as means to an end, but as ends in themselves. "What were the first signs of my love?" Pozdnyshev asks himself, for his interlocutor's benefit. "They were that I abandoned myself to animal excesses, not only quite unashamedly, but even taking pride in the fact that it was possible for me to indulge in them, without ever once taking thought for her spiritual or even her physical wellbeing" (56).

In the second movement of Tolstoy's sonata – the adagio as it were – the moral contrast between animals and humans vacillates so wildly that it often collapses into conflation (if only to better illustrate the potential ethical distance in some more enlightened epoch). Almost like a mantra, variations of the phrase "pigsty existence" (or "swinish behavior" in some translations) are bitterly spit out of Pozdnyshev's mouth, only washed back by the much-too-strong tea which he offers our almost speechless narrator. Love, for the murderer, is thus nothing but lust made presentable, the public face of animalistic animosity. "I was regularly affected by bouts of animosity that used to correspond to the bouts of what we called 'love,'" confesses Pozdnyshev. "We didn't realize it then, but this 'love' and animosity were just two sides of the same coin, the same animal feeling" (78). Furthermore, love, in practice, is "just a sordid matter that degrades us to the level of pigs, something it's vile and embarrassing to remember and talk about. After all, nature didn't make it vile and embarrassing for no reason" (56). Again, the confusion concerning that ur-referent "nature" comes to the fore. What is more shameful – to act naturally like an animal, or to pervert nature to the degree that most humans do, turning raw sex into a slow-cooked sensuality (as we have seen, both monkeys and Parisians do)? What precisely is to blame for this sorry situation of lust: nature or artifice, especially given that the counter-phrase "instrument of pleasure" weaves its way through the same narrative? Men are simians in suits, just as women are exquisitely calibrated, but highly immoral, stimulation machines.

Humanity is thus described as "the filthy king of nature" – a liminal sovereign who refuses to abide by the laws of his own kingdom:

> The animals seem to know that their offspring assure the continuation of the species, and they stick to certain laws in this regard. It's only man who doesn't know these laws, and doesn't want to know them [...] You'll notice that the animals copulate with one another only when it's possible for them to produce offspring [not true, of course, ethologically speaking]; but the filthy king of nature will do it any time, just so long as it gives him pleasure. (60)

Love, this putative "pearl of creation," is little more than the elevation of the "monkey pastime" of carnal relations. The inconsistencies are enough to invoke vertigo at this stage in the tale. Animals are obliged to represent both good morals and disgusting habits. And yet it is only man who would continue to have sex after conception, something Pozdnyshev remembers with a shudder, given that he did not choose to suspend his connubial rights during any of his wife's five pregnancies. Nevertheless, this exclusively human transgression of natural laws is described once more as "pig-like."

exhibit 6: the mutated mother hen

Which leads us to the sixth inhabitant of the bestiary: the mutated mother hen. Pozdnyshev describes the situation as follows:

> And so for the woman there are really only two ways out: one is to turn herself into a freak of nature, to destroy or attempt to destroy in herself her faculty of being a woman – a mother, in other words – so that the man can continue to take his pleasure without interruption; the other isn't really a way out at all, just a simple, gross and direct violation of the laws of nature, one that's practised in all so-called "decent" families. In other words, the woman has to go against her nature and be expectant mother, wet-nurse and mistress all at the same time; she has to be what no animal would ever lower itself to be. (58)

Our protagonist thus anticipates Heidegger, and other thinkers of the "creaturely life" of mankind, in believing that "Man can sink lower than the animal."[15] Contraception is considered just such a fall, since "with the help of those shark doctors" Pozdnyshev's wife is able to prevent pregnancy, and thus becomes "a complete whore," descending below the level of an animal to mere material object (for it was only procreation which gave any kind of alibi to the sleazy status quo) (63). And yet, anticipating the stinging pathos of the climax of the story, Pozdnyshev also sees his pregnant wife and his children as embodying a certain animal or

natural ideal, at least during quick glimpses in the rear-view mirror of hindsight. "For these creatures," he notes, meaning the fruit of their union, "she felt a passionate animal devotion."

The poor woman, however, is trapped between the two poles which constitute the human condition – the animal and the angel – at least when it comes to her maternal side.[16] In a sobering reminder that today's much-lamented "helicopter parenting" was not a recent invention, Pozdnyshev's retrospective compassion clearly emerges when he thinks back to his wife's anxieties over the children's health and habits. As he explains:

> she did not, however, have what the animals have – an absence of reason and imagination. The hen isn't afraid of what may happen to her chick, knows nothing of all the diseases that may attack it, or all those remedies human beings imagine will save them from sickness and death [...] If the chick dies, she doesn't ask herself why it has died, or where it's gone, she merely clucks for a while, then stops, and goes on living as before. (72)

What is usually considered a metaphysical lack on the part of the animal, is here – exceptionally – figured as faculty. But the deceased was trapped in a double-bind, by both her species-being and her individual character: "After all, if she'd really been an animal, she wouldn't have suffered like that; if she'd really been a human being, she'd have believed in God, and she'd have said and thought what the peasant women say: 'The Lord gave, and the Lord hath taken away'" (74).

exhibits 7, 8, 9 and 10: the shape-shifting rook, the bitch-in-heat, the cuckold, and the beast-with-two-backs

Moving now a little faster through the exhibits (in order to allow more time to linger just up ahead), we come to the seventh specimen in our curated zoological garden: a sinister shape-shifter who recognizes and responds to "the woman" buried beneath the duties and concerns of motherhood. He is an interloper who sees not a beast-of-burden but an alluring quasi-feline femininity. He is the snake in the grass, the intruder, the rival, who flatters Pozdnyshev's wife and executes his rather automatic program of seduction, optimized by years of default debauchery. Our protagonist takes an instant dislike to this man, for little other reason than he is a man, hovering around his now radiant wife, suddenly freed from the shackles of cyclical biology. (In passing, on your left, you can just glimpse our eighth exhibit, the bitch-in-heat – of course, she is one of the least understood of creatures, since she does not have a voice in Tolstoy's menagerie.) However, were we forced to settle on *one* phenotype for the character of "the other man," for the sake of taxonomic convenience, then we would be forced to choose between the cock – an almost universal phallic symbol – and the rook; a bird which, accurately or not, has traditionally been a symbol of the incorrigible theft of women-folk, and for the willful fouling of nests. Given the specific resonance of "home-wrecking" of the rook, we shall settle on this, a species notorious for creating the ninth of our cast of creatures: the cuckold.[17]

Slavoj Žižek makes the compelling point that a paranoid husband, who thinks his wife is having an affair, is *no less paranoid* if his wife *is* actually having an affair. This observation has the advantage of avoiding the reflex logic of bourgeois censure, and helps us better identify the psychosocial mechanisms which both muffle and amplify jealousy, in heavily coded terms (the classic triple-play of race, class, gender, but also – as I have been arguing – species). The literature on jealousy is vast, and it would take an entire book even to begin addressing such a complex and significant phenomenon. But as I suggested at the very beginning of this piece, jealousy evolved into a new genus in the nineteenth century, because of the radical restructuring of the meta-categories I just mentioned, and the creation of more fluid commerce between them. As the uncharted internal territory of the psyche is charted by Charcot and then in more detail by Freud, a profound unease emerges, coupled

with excitement, at the implications of this completely unprecedented self-reflexivity in terms of subjectivity itself. In simpler terms, people began to explicitly reflect on their own self-constitution, and the perceived threats to that very same self-constitution. The self was exposed as a fragile and always already jealous ego. Indeed, since Hegel, identity is defined negatively and absolutely as a struggle of recognition,

Jealousy becomes, in and of itself, a *problem* – especially for a class that considers itself the torch-bearers of a more reasonable and enlightened age, unencumbered by the kinds of brutal Old Testament codes of ownership inscribed in the *Domostroy* (mentioned twice at the beginning of the book, and personified by the reactionary old man). Shakespeare's *Othello* literally set the stage for a less allegorical and more psychological exploration of jealousy by novelists who were, after all, almost exclusively members of this new historical class. And they realized that this most violent and unstable of affects threatened to undo all those "techniques of the self" which promised to make society such a functionally civilized experience. Thus we see in Maupassant's last book, translated recently as *Alien Hearts* (and published in the same twelve-month period as *The Kreutzer Sonata*), a protagonist who is a proto-Proustian Swann: surely the ultimate, and still the most lacerating, portrait of male sexual jealousy.[18]

In Maupassant's story, the protagonist, Mariolle, labors under the impression that "he had gained what he had always dreamed of, always sought: complete possession of a woman he loved." However, "[s]uch completeness is not of this world." And so, "he would never embrace for his very own the ever-extending surface of this woman who belonged to everyone" (92). The other gender begins in this period to unfold itself in a form of ontological origami, defying any attempt to contain her in oppressively Euclidian romantic or domestic spaces. As Mariolle's lover, Madame de Burne, begins to speculate silently, and a little sadly, "Could it be that her fine delicate flesh, so exceptionally aristocratic and refined, harbored unknown shames, shames of a superior and sacred animal, shames still unknown to her modern soul?" (113). A new cultural phenomenon is recognized – a women's ego, with its own sovereign imperatives and impulses – and is ambivalently condemned by the masculine establishment as pure, frivolous selfishness, while *also* being celebrated as the obscure cathected target of luminous fascination. That obscure object of desire indeed! Hence the million dollar Freudian question – "what does woman want?" – is asked for the first time in earnest: a question which to this day remains unanswered (at least if the pathologically jealous Mel Gibson is anything to go by).[19] However, the reference to extra-humanity persists – "a superior and sacred animal" – suggesting that any tentative answer should be sought on the hinge between the theological and the evolutionary. (A hinge recently lubricated by the rather viscous intelligence of Giorgio Agamben, of which more in the concluding section of this essay.)

All of which makes it all the more surprising that Tolstoy expressed disdain for Maupassant's novel, on the grounds that none of the characters were tagged with moral legibility – surely a charge one could make of *The Kreutzer Sonata* as well (unless, conversely, one wants to make the case that Tolstoy's novella is overly didactic in this regard).[20] It would take nearly half a century, two world wars, and another Russian before jealousy would be depicted with such feverish, penetrating, and unflinching detail – Vladimir Nabokov's *Lolita*. As detailed in the introduction, Nabokov's narrator, Humbert Humbert, is forced to call upon the mundane materials of his pen to re-capture the object of his desires (and thus the catalyst of his being: according to the same metaphysical logic that a guitar string does not really exist until it is strummed according to a seemingly pre-destined chord). Not for his mortal hands are the "aurochs and angels" of more divine distillers of exceptional experience.

And so, for all the abject animosity that jealousy creates, the question still beckons, in terms of its fraternity with animals or animality. In other words, do animals experience "jealousy" as such? Certainly not according to

Heidegger, for whom animals are incapable of the "as such." So is it more a case of natural competition and selection? I leave this speculation to the ethological experts, but the trope is established: competition and instinctual envy is natural. Jealousy is cultural. The green-eyed monster is thus not a primal creature emerging from the slime and ooze of reptilian resentment, but a modern symptom of specifically human melancholia. It can be found at that smoking interface of agonistic traction where the rubber of the ego meets the road of the superego. What is more, there is a perverse *jouissance* to be found deep within its mental and emotional anguish.[21] As Deleuze notes, "there is something sublime in the jealous man's memory" (52). As men finally acknowledge that women are their own autonomous beings, female attentions can no longer be taken for granted, or simply purchased. Or rather, their attentions *can* be purchased in various currencies, and under the right conditions, but not their hearts or souls. Hence the title of Maupassant's book, and hence Pozdnyshev's childish tantrum that: "What was really so horrible was that I felt I had a complete and inalienable right to her body, as if it were my own, yet at the same time I felt that I wasn't the master of this body, that it didn't belong to me, that she could do with it whatever she pleased, and that what she wanted to do with it wasn't what I wanted" (124–25). And it is this profound, almost metaphysical, insecurity which inflames passion (for the libido responds positively, seemingly perversely, to the possibility or actuality of rejection).[22]

One of the most vivid illustrations of this sado-masochistic self-infliction and self-indulgence is provided by Luis Buñuel's classic film *That Obscure Object of Desire*, in which the love-lust interest literally vacillates between two incarnations, and the jealous male lover is obliged to watch her make love to another (younger) man, while trapped on the other side of a locked and barred gate. The man is literally imprisoned by his own impotent and enforced voyeurism, and he is forced to watch his ego's worst-case scenario, something he flees from at first, but then returns to, compelled by the *frisson* of his libido. And so he is almost disappointed to find that the act has finished without his bearing agonistic witness, or perhaps may have only been simulated, as the woman soon insists (although her statements are far from reliable throughout the film). As Niklas Luhmann says, the lover "enjoys" equally the sweet fruit and the bitter, for the only consistent demand is intensity, as an antidote to libidinal entropy.[23]

Jealousy is thus less about an embodied rival than the very structure and geometry of modern romantic love.[24] If all love is mediated or triangular, as René Girard famously demonstrates, then jealousy is inevitable (at least as long as "possession" is a decisive motif – the jealous person is possessed by the knowledge that it is impossible to truly possess).[25] That is to say, no couple avoids *being*, let alone *becoming*, extra-dyadic (as the psychologists call it). Pozdnyshev seems aware of this dilemma when it comes to the passively aggressive third term in his marriage. Like Anna Karenina's husband, Alexei Alexandrovich Karenin, he at first tries to transcend his negative feelings, aware that, if cultivated, they will bloom fatal flowers.[26] But in contrast to the situation in Tolstoy's most celebrated novel, there is no fork-in-the-road where personal enlightenment might conceivably win out against crude self-interest and flattery.[27] The tension between public face and private pragmatics is narrated in *The Kreutzer Sonata* with little room for generous or altruistic considerations of anyone else.[28] A traditional premium on domestic honor wins out against personal development and inter-personal compassion, thanks to what Žižek calls, by way of Petrarch, "the plague of fantasies." These are the claustrophobic, crowding images that assault the jealous person, creating the kindling for the green flames to lick ever higher. Pozdnyshev shivers as he recalls: "my imagination: it began to paint for me, in the most lurid fashion, a rapid sequence of pictures which inflamed my jealousy [...] a kind of strange, drunken enjoyment of my own hurt pride" (120–21). The main character in this nightmare phantasmagoria – the prime exhibit in the human freak show – is, of course, the next and tenth specimen in our menagerie: the beast with two backs. And as with other freak shows, the lure is a combination of disgust and curiosity.

However, as with the bitch-in-heat, there is no need to dwell on this part of our tour, as we all know what the dual-backed beast looks like, and we all know that its structuring absence is the pivot around which the tale of revenge takes place. Even long after the fateful deed, Pozdnyshev probes his memory for traces of his rival, as one's tongue compulsively searches for the most painful part of a tooth. The husband notes that his potential rival "had a particularly well-developed posterior, as women have, or as Hottentots are said to have" (87), an elliptical reference to a virile, almost equine, animality. And yet there was nothing particular about this fellow, who – like the wife – goes without a name, clothed only in pronouns; nothing specific to fascinate the lady of the house: "If it hadn't been him, it would have been someone else, it had to happen" (86). The inevitability of infidelity has been flagged all along, nurtured since birth by the social matrix. The tragedy of this novella is that the anti-hero chooses murder over divorce, and that he spirals into white-knuckled grasping and lashing out, rather than learning that painful lesson of letting go, and letting be. "Are there not lovers," asks Maupassant, in an observation seemingly tailor-made for *The Kreutzer Sonata*, "who are retained for better or worse, with resignation, out of fear of *the next*?" (148).

And yet there is one skill that marks this gentleman out from the throng of would-be lovers for the wife who is chafing at the bit, and that is his musical proficiency, especially on the violin. The same culture which officially forbids adultery ensures the maximum opportunities for such, via sanctioned social intercourse between the sexes. And as Pozdnyshev observes, the playing of music, *a deux*, is one of the most commonly exploited scenarios – a case of bourgeois morals thwarted, or at least tested, by bourgeois rituals. Which leads us to the eleventh entrant in Tolstoy's bestiary: the siren, aka music itself.

exhibit 11: the siren

Pozdnyshev stretches the credulity of the listener when he insists that his jealousy was not the cause of his psychological and juridical torment, but rather the effect. As he explains to the narrator, the culprit is really his rival. "He and his music were the real cause of it all. At my trial the whole thing was made to look as though it had been caused by jealousy. Nothing could have been further from the truth. I'm not saying jealousy didn't play any part at all, mind – it did, but it wasn't the most important thing" (87–88). Were we to subtract one element from the cascading series of events, however, then clearly Pozdnyshev's jealousy is more decisive and destructive than the aural ghost of Beethoven. What is more, it was the husband's jealousy which – at least according to his own account – *created* the ideal conditions for their retroactive justification: "I saw that right from the first meeting her eyes began to shine in a peculiar way and that, probably as a result of my jealousy, there was immediately established between them a kind of electric current which seemed to give their faces the same expression" (96). (As the literary critic Dorothy Green quotes in a different translation, "From the first moment his eyes met my wife's I saw that the animal in each of them, regardless of all conditions of their position and of society asked 'May I?' and answered 'Oh yes, certainly!'")

Nevertheless, it is through the siren song of music that the wife and her lover ensure the possibility of sensual consummation. (As the debauched master of innuendo Maurice Chevalier exclaims in *The Smiling Lieutenant* – "I *love* chamber music.") For Pozdnyshev, in contrast, it is a "fearful medium," for the very same reason that the well-to-do skirt-chasers love it, because it is "the most refined form of sensual lust" (117). Tolstoy himself was

> an accomplished pianist and made up his mind at one stage to become a great musician and composer. He was carried away by the combination of sounds, and attempted to formulate his own theory of harmony, under the title *The Fundamentals of Music and Rules for its Study*. As far as composers were concerned, Beethoven seems to have inspired in him a kind of love–hate relationship; perhaps he resented Beethoven's power to carry him out of himself.[29] (Green 441–42)

Such resentment is expressed by the murderer as well, when obliged to reflect on the piece of music that makes up the title of the novella:

> Do you know its first movement, the presto? [...] Ah! It's a fearful thing, that sonata. Especially that movement. And music in general's a fearful thing. What is it? I don't know [...] Music makes me forget myself, my true condition, it carries me off into another state of being, one that isn't my own [...] the effect produced by music is similar to that produced by yawning or laughter: I may not be sleepy, but I yawn if I see someone else yawning, I may have no reason for laughing, but I laugh if I see someone else laughing. (110–11)

The power which this particular piece of music has – the intangible force and inexpressible significance which it represents – anticipates Vinteuil's melodic phrase in Proust, which becomes a hook buried in Swann's mind, just as Beethoven places a barb in Pozdnyshev's.[30] As George Steiner (one of the great commentators on Tolstoy, of course) observes: "Because music is so immediately inwoven with changes in the shapes of time, the development of Beethoven's tempi, of the driving pulse in his symphonic and chamber music during the relevant years, is of extraordinary historical and psychological interest" (12). It is this self-same "driving pulse" which Pozdnyshev blames on initiating the fatal sequence of thoughts and emotions – the "cause and affect," as it were – which would lead to our penultimate exhibit in Tolstoy's zoo: the wild, murderous beast.

exhibit 12: the wild murderous beast

Pozdnyshev confronts the agonistic dialectic we have been exploring throughout, between animosity and animality, most searingly at the climax of his tale: a climax he painfully relives in the retelling (and who can say how many times he regales his fellow passengers with this grisly confession?). "I'm seized with horror whenever I think of the wild beast that lived in me during that time," he admits (100–01); ashamed also of the fact that, "In the last days our quarrels became terrifying; they were particularly shattering because they alternated with bouts of animal sensuality" (88). The more he suspects the affair, the more his senses – his nonhuman senses – are inflamed by the possibility.[31] The "animosity" (*zhivotnoe*) which overwhelms him reaches the level of a linguistic tic or mantra

> I was seized by a feeling of animosity towards her more terrible than any I'd ever experienced before.
>
> For the first time I felt a desire to give my animosity physical expression. I leapt to my feet and went up to her; I remember that at the very moment I got up I became aware of my animosity and asked myself whether it was a good thing for me to abandon myself to this feeling, and then told myself that it was a good thing, that it would give her a fright; then immediately, instead of fighting off my animosity, I began to fan it up in myself even further, rejoicing in its steadily increasing blaze within me.
>
> "Go, or I'll kill you!" I shouted suddenly, going up to her and seizing her by the arm, consciously exaggerating the level of animosity in my voice. (106)

Five repetitions of the word in barely more than a paragraph! What is more, Pozdnyshev's wife remains "a mystery, just as she's always been, just as she'll always be. I don't know her. I only know her as an animal. And nothing can or should hold an animal back." The two forces feed on each other with a deepening hunger.

After leaving his wife alone on official business, Pozdnyshev's paranoia draws him back to home and hearth, only to be confirmed in such sickening circumstances that time itself seems to be stretched into a substance of pure nausea. As feared, the rook is there, befouling the nest of the cuckold. Despite the fine

clothing and the expensive China, Pozdnyshev walks in on a beastly scene – his wife sharing an intimate dinner with his lover – so that he only sees something inhuman. "No, she isn't human, she's a bitch, a repulsive bitch!" (130).

The scene in which the crime of passion takes place, by way of a knife between the ribs, is a masterpiece of unsettling representation: the equivalent of catching lightning in a bottle, in terms of the technopoetics of capture. Here a tragic "event" is logged microsecond by microsecond, in forensic detail, but fleshed out via the "unreal time" of psychological distress (as Tolstoy also does with such consummate skill for Anna Karenina's suicide). Pozdnyshev's wife does not die immediately though – her mortal flame continues to flicker to better confront him with the irreversibility of his actions. Even then, as her life flows silently out of her body, the madman wrenchingly recalls, "I saw displayed on her face the same inveterate look of cold, animal hatred I knew so well" (142). He continues:

> I looked at the children, at her battered face with its bruises, and for the first time I forgot about myself, about my marital rights and my injured pride; for the first time I saw her as a human being [...] I realized that I'd killed her, that it was all my doing that from a warm, moving, living creature she'd been transformed into a cold, immobile, waxen one, and that there was no way of setting this to rights, not ever, not anywhere, not by any means. (142–44)

And here lies the supreme irony of Tolstoy's tale; Pozdnyshev cannot fully register his wife's humanity – that is her monadic autonomy – until the very moment she is reduced to "animalistic" bare life. Or rather, bare lifelessness. And, of course, at that same moment it is too late. In the agonizing hour between the puncturing of her organs, and the breath leaving her body forever, the adulteress exists in that fraught zone Eric Santner calls "creaturely life": an abject space in which the inherent pathos of the human condition is rendered animal, only to ontologically enhance the Fall from its assumed and exceptional dignity.

Again, the irony is that our nebulous "humanity" is touched most to the quick when our humanness is swiftly compromised, evacuated, or stolen.[32] Pozdnyshev appreciates his wife's personhood only when he holds a bruised, virtual carcass – a revelation denied him by the blinders of jealousy; that "fatal energy" (95) which seeks only revenge, because it cannot see beyond the horizon of the ego. As Jean Baudrillard so accurately puts it:

> To love someone is to isolate him from the world, wipe out every trace of him, dispossess him of his shadow, drag him into a murderous future. It is to circle around the other like a dead star and absorb him into a black light. Everything is gambled on an exorbitant demand for the exclusivity of a human being, whoever it may be. This is doubtless what makes it a passion: its object is interiorized as an ideal end, and we know that the only ideal object is a dead one. (105)

Pozdnyshev's remorse stems directly from reaching this ideal, and the reader wishes that he had the foresight and wisdom of Tolstoy's own wife, Sofia, who at least anticipated the agony which follows swiftly on the heels of the ecstasy of passionate homicide. "If I could kill him and create a new person exactly the same as he is now," she wrote, only a matter of months after their wedding, "I would do so happily" (qtd in Meek).

exhibit 13: the empty shell of a man – by way of summary and conclusion

Which brings us to the final, unlucky thirteenth exhibit in the bestiary: the haunted creature, the husk of a human being. As readers – as visitors to Tolstoy's zoo – we can do little more than pity the poor fellow, or at least find him pitiful. Whether we as proxied interlocutors forgive Pozdnyshev after his full confession is not really the point. Nor should we measure our own reactions against "the moral of the story," as painstakingly detailed by the author himself – or at least, not only. For as I have argued, there is a veritable jungle of animal types

which cannot be completely domesticated by Tolstoy's pen. Obliged to personify in reverse, these creatures suggest a struggle far more widespread than between husband and wife, man and woman, but one identified by one of the author's most controversial and celebrated contemporaries.[33] For Nietzsche, humanity was little more than the animal that had learned to make promises; the marriage vow being one of the most iconic. Indeed, *The Genealogy of Morals* alludes to the rather terrifying set of technologies assembled by men as pre-emptive and punitive mnemonics, so that this rather frail and forgetful creature remembers to remember. Hence morality itself, as the most splendid and suffocating architecture. For men like Tolstoy, morality is a cathedral, reaching for the heavens. For those like Nietzsche, it is nothing more than a cage, preventing man from transcending himself toward an empowering animality. Tolstoy's obsessive cataloguing of animals in *The Kreutzer Sonata* could be read as his attempt – conscious or unconscious – to grapple with Nietzsche's challenge (itself a provocation enabled by Darwin).

As such, this novella represents a particularly fraught cog in the discursive device Agamben calls "the anthropological machine," that is, a comprehensive engine designed *by* humanity and *for* humanity, whose primary function is to sort "the human" (as abstract privileged subject) from the nonhuman (an equally abstract object). In terms simplified to a degree that may give Agamben himself a migraine, the anthropological machine is all those cultural artifacts which combine into a vanity mirror for our species, reflecting and regulating our assumed superior status. Agamben thus considers the trajectory of humanity to be at once located on a continuum with other animals – after all, Linnaeus himself could not offer any unique identifying features – *and* exceptional in its self-recognition as different. In other words, it is only through our delusional narcissism that "we" stand apart as a modern, secular Adam. And yet, miraculously, this delusion (aka, anthropocentrism) seems to be enough to forge an ontological difference, given all the mediating devices engineered and employed to maintain that very distinction. (This is where and why Agamben can still be accused of a latent humanism, with tinges of the theological.)

In any case, Agamben shares Nietzsche's fascination with a post-historical transcendence of ourselves, in which we would be ultimately reconciled with our animal natures, and all the freer for it. This would not be a return to a bestial state but an inclusive fusion of soma and spirit, no longer troubled by the agonistic detour of culture (as defined against and in contrast to nature). It would be a last supper as illustrated in the thirteenth-century Hebrew Bible, with "the righteous" depicted with animal heads, rather than human faces (1–3). For Tolstoy, this is anathema. The spiritual superiority of humankind can only be understood in relief to an animalistic background. And as long as men and women lapse back into the sensual realm, their bestiality will create a dangerous animosity.

Let us recap, then, our zoological itinerary, cage by cage. First there was the green-eyed monster himself, gnashing his sharp, greedy teeth, and possessed with the impossibility of possession. Then the fornicator, who indulges in debauchery at the expense of his consorts. Followed by the doctor-reptile, who encourages such irresponsible behavior, under the relatively novel alibi of public health. Remember too, the Venus fly-trap – the female response to such ravishments, determined to parry her social disadvantage into an economic advantage. Together, the fornicator and the Venus fly-trap morph into the porcine couple, in the nuptial chamber, using each other for pleasure only, as means rather than ends. Which in turn leads to the mutated mother hen, who first violates herself, and then her child, by continuing to have sex after conception. Eventually she learns – once more with the aid of the reptile-doctors – to avoid conception altogether (one of many noted inconsistencies or ironies in Pozdnyshev's account, whereby human science paradoxically and perversely encourages a heightened animal state). In turn, sterility creates the bitch-in-heat, not for procreation, however, but for a wicked kind of gratification.

With the active complicity of the siren (music), she is subsequently easily seduced by the cock or rook, whose own sensual delights are at the expense of the cuckold. Upon gaining indirect carnal knowledge of his wife's infidelity, the cuckold is confronted with the mocking specter of the beast-with-two-backs; an image so infuriating that it can lead to a Hulk-like transformation, into a murderous wild beast. The storm of passion passes, however, as soon as the irreversible deed is done, leaving only the haunted shell of a man.

These are the figures deployed in Tolstoy's story – sometimes explicitly, sometimes through allusion – to describe the ultimately *in*human dispensation of love. Moreover, Pozdnyshev's tale illustrates how jealousy can function as an efficient lubricant for the smooth operation of the anthropological machine, thereby fueling all the undesirable products of species-based chauvinism. Tolstoy was correct in insisting that humankind will forever languish in spiritual limbo as long as it hitches its wagon to institutional matrimony. He was mistaken, however, in attempting to locate humanity in monogamous chastity (and not only because he was incapable of practicing what he preached). Sex is not the problem, but rather the site in which a secular and fleeting form of salvation is possible. At least according to Agamben, sex enables and enacts a "mutual disenchantment" which nevertheless affords "a new and more blessed life, one that is neither animal nor human" (87). Beyond pious, mythologizing pedestals, intimacy reaches the point where it is capable of refusing the potentially fatal fetishism of the individual. As Agamben cryptically puts it:

> These lovers [he is speaking of Titian's painting *Nymph and Shepherd*] have initiated each other into their own lack of mystery as their most intimate secret; they mutually forgive each other and expose their *vanitas* [...] In their fulfillment, the lovers who have lost their mystery contemplate a human nature rendered perfectly inoperative – the inactivity and *desœuvrement* of the human and of the animal as the supreme and unsavable figure of life. (87)

Tolstoy – unlike Titian, and unlike Nietzsche – was incapable of seeing the admirable side of *post*-human, neo-animalistic sexual congress: an admittedly messianic state in which "people" may indeed enjoy each other's bodies and souls without the folly of possessive vanity. Let it be, rather than let it bleed. In contrast, the great Russian author made the ulti-mate human error – assuming the autonomous existence of something called humanity in the first place.

And also in the last place.

notes

1 While the precise origin of the phrase is unknown, it was certainly popularized by Shake-speare's usage in *Othello*.

2 Given its resonance with Pozdnyshev's account, it is worth quoting Nietzsche in full:

> In the three or four civilized countries of Europe, a few centuries of education would suffice to make women into anything we want, even into men – not in the sexual sense, admittedly, but at least in every other sense. Acted upon in this way, they will at some point have assumed all the male virtues and strengths, at the same time, of course, having to assume their weaknesses and vices as part of the bargain: this much, as noted, we can accomplish by force. But how will we endure the intermediate state that this will bring about and that might itself last for a few centuries, during which female follies and injustices, their age-old birthrights, will still assert their supremacy over all that has been newly won and acquired? This will be the time when anger will comprise the essential male affect, anger at the fact that all the arts and sciences have been inundated and clogged with an unprecedented dilettant-ism, that philosophy has been talked to death by bewildering chatter, that politics have become more fantastic and partisan than ever, that society is completely dissolving because the keepers of the old morality have become ridiculous to themselves and are striving to stand outside of morality in every

possible way. For if women had their greatest power in morality, what would they have to grasp in order to regain a comparable ampli-tude of power after having given up morality? (230–31)

3 It is also worth noting the recurring motif of connection between trains and confessional encounters, as practiced also by Maupassant and Buñuel, amongst many others. Anna Karenina herself, of course, is tragically linked with this rather totemic form of transport, singled out by Freud as a privileged stimulator of the libidinal sen-sorium. At one point Pozdnyshev exclaims, "Oh, I'm so afraid, so afraid of railway carriages; I get stricken with horror in them" (122), swallowed up inside this new mechanical beast. He also tells of killing time by visiting "a Jew" in a third class compartment, "the interior of which was spattered with the husks of sunflower seeds" (123), empha-sizing the circus-like aspect of train travel, in which the passengers are little more than animals in transit.

4 The *Domostroy* was a sixteenth-century collec-tion of archconservative domestic rules and guide-lines pertaining to common public and private matters of Russian society. The core values con-tained therein uniformly endorsed obedience and submission to God, Tsar, Church, and Father, especially through modest dress, prayer, the veneration of icons, and charity.

5 While not a daily practice, let us not forget that the practice of putting animals on official trial, and holding them accountable to human law, was only phased out altogether in Europe in the first decade of the twentieth century.

6 Nowadays, rather than simply pulling on the proverbial reins, husbands use a passive-aggressive species of "trust" to try to control the behavior of their spouse.

7 "It's horrible," exclaims Pozdnyshev, "[t]he abyss of error we live in regarding women and our relations with them" (20).

8 In her diary, Sofia Andreyevna reflects on her husband's hypocrisy:

> If only the people who read *The Kreutzer Sonata* so reverently had an inkling of the voluptuous life he leads, and realized it was only this that made him happy and good-

natured, then they would cast this deity from the pedestal where they have placed him! Yet I love him when he is kind and normal and full of human weaknesses. (In Meek n. pag.)

9 According to Pozdnyshev, the very engine of the luxury economy is driven by the libidinal economy of women's covetous desires: "Women are like empresses, keeping nine tenths of the human race in servitude, doing hard labour. And all because they feel they've been humiliated" (40). An observation with perhaps a disconcerting grain of truth even today, if we cast a cold eye on the mediascape's share dedicated to feminized consumables.

10 In a line which no doubt echoes Tolstoy's own thoughts later in his life, Pozdnyshev confesses, in almost a fit of male hysteria, that when he sees a woman in a ball-gown he wants to call the police, "demanding that the hazardous object be confiscated and taken away" (41).

11 This curious tick of Pozdnyshev's — a kind of reverse anthropomorphism — is also in evidence when he paints a portrait of domestic banality, in which man and wife suffer "the sort of conversations I'm convinced animals carry on with one another. 'What's the time? Bedtime. What's for dinner? Where are we going to go? Is there anything in the newspaper? Send for the doctor. Masha's got a sore throat'" (78).

12 For all of Tolstoy's gestures to "human dignity" and related elevated strivings, the question remains concerning the depth of his misanthropy. G.K. Chesterton, for instance, used the occasion of the famous author's eightieth birthday to criticize his attitude toward his own species, noting that

> Tolstoy is not content with pitying humanity for its pains: such as poverty and prisons. He also pities humanity for its pleasures, such as music and patriotism. He weeps at the thought of hatred; but in *The Kreutzer Sonata* he weeps almost as much at the thought of love. He and all the humanitarians pity the joys of men.

Moreover, addressing his target directly, Chesterton adds: "What you dislike is being a man. You are at least next door to hating humanity, for you pity humanity because it is human" (*Illustrated London News* 19 Sept. 1908).

13 Pozdnyshev enquires:

> perhaps you're an evolutionist? The outcome's still the same. In order to defend its interests in its struggle with the other animals, the highest form of animal life — the human race — has to gather itself into a unity, like a swarm of bees, and not reproduce infinitely: like the bees, it must raise sexless individuals, that's to say it must strive for continence. (49)

14 A conclusion taken to a drastic degree by at least one disciple of the master, who sliced off his own penis — the source of animalistic desires, and constant saboteur of human aspirations — before making the pilgrimage to visit Tolstoy. As James Meek notes, in 1909, Sofia wrote in her diary:

> This morning we had a visit from a 30-year-old Romanian who had castrated himself at the age of 18 after reading *The Kreutzer Sonata*. He then took to working on his land — just 19 acres — and was terribly disillusioned today to see that Tolstoy writes one thing but lives in luxury.

Apparently Tolstoy himself wrote in his diary: "An exceedingly interesting man."

15 For a critique of this position, see my article "After the Beep."

16 We see echoes of this inclusion/exclusion dynamic in the widely held contrary position, that "only parents *truly* know what it is to be human."

17 The *OED* explains that

> The origin of the sense is supposed to be found in the cuckoo's habit of laying its egg in another bird's nest; in Ger., *gauch* and *kuckuk*, and in Pr., *cogotz*, were applied to the adulterer as well as the husband of the adulteress, and Littré cites an assertion of the same double use in French; in English, where *cuckold* has never been the name of the bird, we do not find it applied to the adulterer.

18 If the reader will forgive me: literary criticism seems to call for such anachronistic pronouncements of incontrovertible aesthetic fact. Or perhaps I've just been reading too much George Steiner lately.

19 *What Women Want*, dir. Mel Gibson (2000). As I write, Gibson is in legal hot water for threatening his ex-lover over the phone with highly sexualized violence, and thus providing the very voice of hyper-phallic jealous panic. One wonders if this is indeed the voice of Pozdnyshev himself.

20 Two years after Maupassant's death, Tolstoy wrote: "In this last novel the author does not know who is to be loved and who is to be hated, nor does the reader know it, consequently he does not believe in the events described and is not interested in them" (viii). Which only makes it even more ironic that Theodore Roosevelt described *The Kreutzer Sonata* as the product of "a sexual moral pervert" (in Lessing 27).

21 Both the cuckold and the rook are vulnerable to indulging in a particular instance of "the plague of fantasies"; one which represents a variation on the famous fable of Buridan's Ass – the beast of burden that died of hunger, unable to choose between two equally delicious bales of hay. In this instance, when it comes to extra-marital affairs (such as the one in *The Kreutzer Sonata*), neither man has a choice. And yet both are haunted by an impossible emotional alternative: to be the husband, and make love to the beloved while she thinks of another; or be the lover, whose solace-cum-torture is to know that he is in his beloved's mind as she makes love to her husband.

22 Amazingly, as I write this, the blogosphere is a twitter about an alleged new "sex fetish for intellectuals" called Cuckolding. To quote the original article introducing this concept to the wider world, "It's S&M for Ph.D.s […] in which men watch their wives have sex with other guys" – a trend which "is catching on among people with high IQs who revel in the psychological agony." The piece quotes one avid practitioner, "The high point of cuckolding is when your wife says she wants the other guy all the time and never wants you." This fetish is thus presented as a self-reflexive exploration of the libidinal economy, whereby jealousy can be something other than a hostile take-over bid, launched on behalf of an unwise over-investment in the self. As one behavioral expert puts it, jealousy

> is a social construct based on the notion that husbands own their wives, and is thus "much more recent, evolutionarily speaking, than

the competition that turns guys on. That's why it's mostly intellectuals who are into cuckolding: because other guys are crippled by jealousy. They're aroused and upset and don't know why."

The problematic and elitist implications of such statements are obvious enough without metacommentary here. (All quotes from Rufus n. pag.)

23 Luhmann believes that "it is impossible in love to calculate the costs or weigh up the accounts, because both one's profits and one's losses are enjoyed; indeed, they serve to make one aware of love and to keep it alive" (67).

24 Pozdnyshev notes: "In our section of society all husbands are jealous" (102).

25 One cannot help but consider this as a symptom of Lacan's dictum, "There is no sexual relationship." However, in some cases, the more that this metaphysical obstacle is felt, deep in the heart or loins of the lover, the more frenzied and disturbing the attempt to bridge the unbridgeable, to catch the uncatchable. Exhibits A and B for twentieth-century literature would be John Fowles' *The Collector* and Nabokov's *Lolita*.

26 A fascinating, emotionally scarring, postmodern twist on the "tolerant husband" can be found in Lars von Trier's film *Breaking the Waves*. From a certain angle, Bess could be considered a strange love-child of Anna Karenina's – a woman who dies not because of possessive jealousy but quite the opposite, because her (invalid and possibly insane) husband pushes her out into the public sphere, to have sex with as many strange men as possible.

27 Today's "polyamorous" community puts much stock in the concept of *compersion* – a notion originally conceived by the Kerista Commune in San Francisco. Compersion is an extended erotic form of empathy, in which the subject experiences joy via a third term, up to and including the figure usually considered to be a threat or rival. In other words, compersion is the positive and inclusive flip-side of jealousy. (Just as some would say a smiling unicorn is the positive and inclusive form of a snorting, angry stallion.)

28 A fascinating case in Russian life and literature of the same period is Turgenev, who not only enthusiastically introduced Maupassant's writings to Tolstoy but who wrote of unjealous husbands

(cf. *Spring Torrents*) from his literally eccentric position. As one critique notes, "It was a constant refrain of Turgenev that he failed to 'weave himself a nest' in life and had been forced to perch on the edge of strange nests" (Schapiro 197) – most notably the nest inhabited by the Spanish singer Pauline Garcia-Viardot. Clearly Turgenev was not a rook, since he never presumed to poach a woman from another man, and yet neither did he consider wedlock to be an obstacle to living with his true love (*and* her exceedingly tolerant husband).

29 In the same piece, Green makes the interesting claim that

> the reader of this novel, should be in a very real sense a listener. We need to remind ourselves again that Tolstoy conceived his story with the living voice of an actor in his mind. That is, it was written, as the sonata was, for an instrument. (447)

30 See the illuminating comments on "the refrain" in Deleuze and Guattari's *A Thousand Plateaus* (chapter 11), and on Vinteuil's phrase more specifically in Deleuze's *Proust and Signs*.

31 No doubt an evolutionary biologist, or even sociobiologist, would read such passion as the cultural froth of genetic imperatives. But as useful and sobering as such accounts are in reminding us of our mammalian heritage, they are almost universally reductive when it comes to interpreting and contextualizing human behavior. Psychology, affect, aesthetics, and other "cultural" factors are ignored, or, at best, alluded to as inessential contingencies (cf. Jonason et al.'s article "Positioning the Booty-Call Relationship on the Spectrum of Relationships"). A point I make not to preserve the exceptionalism of humans in this regard but rather to identify as underdeveloped influences in general within such fields. Of course, the humanities have been guilty of over-emphasizing the role of finer forces, and the more sharing of conceptual lenses we can have between the humanities and the sciences, the more comprehensive and revealing will be the readings.

32 The vast and growing literature around this is usually framed by questions of biopolitics (cf. the vast bibliography issuing from Agamben's rendering of Foucault's concept, especially in *Homo Sacer*).

33 For an insightful, and long overdue, reading of Nietzsche's nuanced relationship to the animal kingdom, look no further than Vanessa Lemm's book *Nietzsche's Animal Philosophy*.

bibliography

Agamben, Giorgio. *The Open: Man and Animal.* Trans. Kevin Attell. Stanford: Stanford UP, 2004. Print.

Baudrillard, Jean. *Fatal Strategies.* Trans. Philip Beitchman and W.G.J. Niesluchowski. Ed. Jim Fleming. New York: Semiotext(e), 1990. Print.

Buñuel, Luis, dir. *That Obscure Object of Desire.* Greenwich Film Productions, 1977. Film.

Deleuze, Gilles. *Proust and Signs.* Trans. Richard Howard. Minneapolis: U of Minnesota P, 2000. Print.

Girard, René. *Deceit, Desire and the Novel.* Trans. Y. Freccero. Baltimore and London: Johns Hopkins UP, 1988. Print.

Goethe, Johann Wolfgang von. *Maxims and Reflections.* Trans. Elisabeth Stopp. London: Penguin, 1998. Print.

Green, Dorothy. "The Kreutzer Sonata." *Tolstoy's Short Fiction: A Norton Critical Edition.* Ed. Michael R. Katz. Austin: U of Texas P, 1991. Print.

Jonason, Peter K., Norman P. Li, and Jessica Richardson. "Positioning the Booty-Call Relationship on the Spectrum of Relationships: Sexual but More Emotional than One-Night Stands." *Journal of Sex Research* (July 2010). Web. 7 Aug. 2010. <http://dx.doi.org/10.1080/00224499.2010.497984>.

Lemm, Vanessa. *Nietzsche's Animal Philosophy: Culture, Politics, and the Animality of the Human Being.* New York: Fordham UP, 2009. Print.

Lessing, Doris. "On Tolstoy." *Time Bites: Views and Reviews.* New York: Harper, 2004. 27–41. Print.

Lubitsch, Ernst, dir. *The Smiling Lieutenant.* Paramount Pictures, 1931. Film.

Luhmann, Niklas. *Love as Passion: The Codification of Intimacy.* Trans. Jeremy Gaines and Doris L. Jones. Stanford: Stanford UP, 1998. Print.

Maupassant, Guy de. *Alien Hearts.* Trans. Richard Howard. New York: New York Review of Books, 2009. Print.

Meek, James. "Some Wild Creature." *London Review of Books* 32.14 (2010): 3–8. Web. 7 Aug. 2010. <http://www.lrb.co.uk/v32/n14/james-meek/some-wild-creature>.

Nabokov, Vladimir. *The Annotated Lolita*. Ed. A. Appel, Jr. New York: Vintage, 1991. Print.

Nietzsche, Friedrich. *Human, All Too Human*. Trans. Gary Handwerk. Stanford: Stanford UP, 1995. Print.

Pettman, Dominic. "After the Beep: Answering Machines and Creaturely Life." *boundary 2* 37.2 (2010): 133–53. Print.

Pettman, Dominic. *Human Error: Species-Being and Media Machines*. Minneapolis: U of Minnesota P, 2011. Print.

Proust, Marcel. *In Search of Lost Time*. Trans. D.J. Enright, Terence Kilmartin, and C.K. Scott Moncrieff. New York: Modern Library, 2003. Print.

Rufus, Anneli. "Cuckolding: The Sex Fetish for Intellectuals." *The Daily Beast* 29 July 2010. Web. 5 Aug. 2010. <http://www.thedailybeast.com/blogs-and-stories/2010-07-29/cuckolding-the-sex-fetish-for-intellectuals/>.

Santner, Eric L. *On Creaturely Life: Rilke, Benjamin, Sebald*. Chicago: U of Chicago P, 2006. Print.

Schapiro, Leonard. "Critical Essay – *Spring Torrents*: Its Place and Significance in the Work of Ivan Sergeyevich Turgenev." *Spring Torrents*. By Ivan Sergeyevich Turgenev. London: Penguin, 1980. 183–239. Print.

Steiner, George. *In Bluebeard's Castle: Some Notes towards the Redefinition of Culture*. New Haven: Yale UP, 1971. Print.

Tolstoy, Leo. *Anna Karenina*. Trans. Richard Pevear and Larissa Volokhonsky. London: Penguin, 2006. Print.

Tolstoy, Leo. *The Kreutzer Sonata*. Trans. David McDuff. London: Penguin, 2007. Print.

Žižek, Slavoj. *The Plague of Fantasies*. London and New York: Verso, 1997. Print.

the anthropocentric frame of screen animals

When animals appear as central figures in mainstream films, they tend to stand in for none other than human characters or nothing but unruly antagonistic nature. In Hollywood cinema, these extreme cases are salient in two typical genres: animation and disaster. A lion prince in *The Lion King* (Roger Allers and Rob Minkoff 1994), tricked by his uncle into believing he killed his father, flees the kingdom, but after years of exile he returns home to overthrow the usurper and retrieve his royal identity. What is staged in this Disney animation is not a National Geographic on wildlife but a human drama newly mixing the old motifs of Oedipus and Hamlet in the character-driven, goal-oriented classical Hollywood narrative. On the contrary, the SF adventure *Jurassic Park* (Steven Spielberg 1993) shows a utopian theme park with biotechnologically created dinosaurs turning into a catastrophic dystopia by accident, in an instant. Animals may look "beautiful" when put in a touristic zoo under human control, but their potentially insurmountable power is in nature "dynamically sublime" in Kant's terms, always ready to manifest itself as dangerous monstrosity that can run amok, terrifying us and making us feel powerless.

These two oppositional modes of animal representation, however, work in the same anthropocentric paradigm in which the notion of nature, the animal world, could not come into being without its insertion into the cultural dichotomy of nature and culture. Nature did not preexist culture in that its idea was not born until culture named and incorporated it

seung-hoon jeong

A GLOBAL CINEMATIC ZONE OF ANIMAL AND TECHNOLOGY

into the conceptual frame of what humans believe as reality.[1] Only within this frame does nature appear to be the opposite of our lifeworld, while the frame itself remains cultural. Therefore, animals exist as the Animal only and always as viewed by, and related to, the Human. We immediately recognize animal allegories for human characteristics, good or evil, brave or cowardly, generous or greedy, and so on (thus animal characters are inherently civilized); otherwise we consider animals to be either domestic and helpful or untamable and harmful (thus the prevailing "pet or pest" binary persists). Our binary attitude to them, at least in our civilized safety without sublime threat from wild animals, is then sentimental or brutal, "sometimes aglow with the welcoming hearth but just as often coldly shutting out the unwanted outsider" (Chaudhuri and Zurkow).[2] The point is not that hospitality can easily change into hostility, but that this Manichean reaction itself deprives animals of their Real that could not be fully symbolized in our reality; or, say, it deprives our reality of room for approaching or encountering their Real. Conversely, the animal Real, even if absolutely aggressive and invincibly destructive to humans, subsists primarily as *being-in-itself*, which we only secondarily view as *being-for-us*, similar or opposed to us.

Then how could we conceive cinematic alternatives that are open to the lost Real? We first need to be aware of a certain self-contradiction in our ideological conception of the animal and nature: the animal is wild in the wilderness, whereas wilderness often connotes nature as the organic, holistic, hippie-spiritualized ground of peace and harmony which has been not only uncontaminated by human civilization but "must be biologically intact and legally protected" as the WILD Foundation states (www.wild.org). This virginal environment valued for moral, cultural and aesthetic reasons remarkably colors the Japanese animation represented by Miyazaki Hayao, whose works have been no less globally consumed and appreciated than those of Disney or Pixar: *Nausicaä of the Valley of the Wind* (1984), *My Neighbor Totoro* (1988), *Princess Mononoke* (1997) and *Spirited Away* (2001), to name a few. Imbued with local animism and spiritualism, the Gaia-evoking Japanese nature he depicts nurtures life and cures wound, exerts magical power and defeats colonial violence. But despite this apparent "green" message, Hayao is evidently not a naïve New Age conservationist appealing mainly to children. *Princess Mononoke*, a unique Asian werewolf film, does not just focus on a human-animal's individual psycho-social struggle in the Western horror format that demonizes either humanity or animality. It rather reveals how the human and the animal are complexly interrelated and how each side is divided as well: beside imperial and samurai forces, a

humanitarian leader builds a self-sustaining commune through the early modern manufacture of firearms; the boars and the apes in nature are destructive and even counterproductive, while the deer-like forest deity *shishigami*, the spirit of nature, stands for neither good nor evil, neither life nor death. Nature is a realm of events simply to be accepted, which appears unfair to the werewolf girl who cries out on her mother wolf's death by humans (Levi 152).[3] In other words, our common sense of wild/erness implies the discrepancy between animal and nature, which actually insinuates the discrepancy within nature itself. Nature is not simply organic in its totality or antagonistic to humans, but deeply antagonistic and even indifferent to itself. One might call this self-destructive inexplicable nature "anti-nature" (*antiphusis*), as Jacques Lacan suggests, insofar as it challenges precisely not the human world but the humanist frame of nature. Anti-nature is the barred Real, not a unified wholeness but a fractured materiality blocked from the symbolic order of smooth linguistic translation and logical intellectual understanding (Johnston 34–37).[4]

If Japanimation exposes this unnatural nature, unlike Hollywood animation, Alfred Hitchcock may be one of the first Hollywood auteurs who introduced a different sort of disaster film with the animal behavior being in no way completely explicable. His famous *Birds* (1963) begins in a pet shop where Melanie buys a pair of lovebirds, but her romantic boat trip with this animal gift to Mitch is cracked by a seagull's sudden hit on her forehead. This small incident is followed by all kinds of bird attacks on the entire seaside town of Bodega Bay in their immeasurable number taking on the "mathematically sublime." We know that this natural violence has a multilayered classical psychoanalytic allegory: the intervention of Mitch's jealous mother, a Hitchcockian superego, in his romance with Melanie; the unconscious attachment as aggressivity inherent in the mirror phase entered by the couple; the punishment of Melanie's active sexuality in phallocentric classical narrative; the uncanny return of the repressed in human civilization with caged animals turning into wild ones; and so forth.[5] We nonetheless find no comprehensively comprehensible answer to the primary question of why birds attack. What is the actual motivation of those peaceful animals' abrupt change into brutal monsters? But the answer lies in the question itself. This incomprehensibility, this unmotivated self-mutation of natural balance is the nature of nature, the anti-natural core of what we take for granted as nature. The first bird's attack on Melanie's euphoric boat in the bay appears, Slavoj Žižek says, as a "Hitchcockian blot" (*Looking Awry* 88–106); a visual smear that triggers the overturn of our picture of reality, the catastrophe of our harmonious ecology of environment – Greek *katastrophē* means overturn. We are helpless in accounting for this intrusion of the Real, and our powerlessness proves nothing but the absurd otherness of the animal.

The mainstream cinema often leaves room for multiple interpretative entry points to this otherness, but again, anthropocentricity is the common hermeneutic matrix of social, political, mythical, religious references or allegories itself. We easily recognize the biblical plagues of locusts in *The Omen* (Richard Donner 1976), the crime investigation through the communication with insects in *Phenomena* (Dario Argento 1985), the bestial eroticism of Western werewolves in *Cat People* (Jacques Tourneur 1942; Paul Schrader 1982) and *Wolf* (Mike Nichols 1994), etc. In the genre cycle of disaster film, *Gremlins* (Joe Dante 1984), for example, where cute eponymous pets turn into malevolently mischievous monsters, sutures the motif of *The Birds* into the 1980s "campy" trend of disaster comedy that parodies former genre films (Feil 31–58). Rather than confronting animality as such, this *kitsch* film transforms the animal as the external ontological Other into internal sociological others of the majority in a human community. Gremlins' mathematically sublime proliferation is then read, ironically or critically, to stand for the growing threat to the mainstream middle-class white America by stereotyped social minorities: the disgusting creatures are black like African

Americans, do whatever like liberals, subvert domestic power relationship like feminists … Likewise, ontological others such as zombies, monsters, ghosts, and vampires have been interpreted as figuring such minorities, including communists, immigrants, foreigners, workers, and even capitalists sucking the blood of the proletariat. In short, animal-like beings on screen cannot help being more or less personified in the frame of cultural studies whose identity politics is built on differences in class, sex, gender, race, ethnicity, and so on, within human societies. And in this aspect they are treated as favorable or unfavorable, our friends or enemies. Thus, again, the two modes of animal representation mentioned at the beginning are intermingled into one: the Animal as/ for the Human.

Undoubtedly, this anthropomorphic tendency works according to what Fredric Jameson calls hermeneutic "depth models": dialectic, psychoanalytic, existential, or semiotic – the hierarchical dichotomy that there is a latent meaning, essence, signified *below* the appearance of manifest signifier (Jameson 12). What matters is the invisible deeper level full of human-oriented meanings and not their animal image. Put differently, however, the absolute difference between human and animal is reduced to relative differences among human-looking animal groups. The "reading" of animals as disguised humans is then at risk of being blind to animals themselves; our vision has a blind spot with regards to their animal being as just seen on screen. More significant than the depth of humanized meanings is the surface of the animal; it is the surface that exhibits the animal as radical difference *above* cultural differences among human-animals. The fundamental conceptual task is therefore to *add* ontological others (including the machine, as we will see) to cultural others while *replacing* the latter in reading films. This *supplement* would help retool established cultural studies while remobilizing identity politics in the contemporary context of the so-called "ethical turn of the political" – to simplify Jacques Rancière's diagnosis, the political conflict among oppositional identity/interest groups in a society now gives way to the ethical one between a global community of harmonious differences as a whole and its exception (Rancière 109–32). The national politics of bourgeoisie vs. proletariat in a country, for instance, becomes less fundamentally decisive than the global antagonism between an encompassing multicultural society and a small band of terrorists. Rancière argues that the ethical turn results in the globalized world's "war on terror" that ends up being indistinct from terror itself in its operation, so he critically points out political side-effects of this turn. Nevertheless, given the direction of ongoing globalization that tolerates ever more diverse class/sexual/racial identities while generating unprecedented global issues beyond national solutions, the place of the remnant of the world system could be conceived of as larger than just terrorism. It is the place of the environment whose catastrophe would impact the entire global village beyond individual political communities. Ontological others of the human call for our attention in this regard, urging us to explore a larger bio-*polis* emerging between, and encompassing both, the human world that becomes ever more globally homogenized and its radically external-immanent environment, natural or technological. The question of how to face this environment requires complexly ethical rather than simply political attitudes, since biopolitics concerns not a new public sphere so much as the condition of any such *polis*, as we will see. We need to review the ethical turn expansively as eco-ontological.

redirecting ethics of the animal

To do so, I look at a series of contemporary films that seem to embody a facet of "global cinema" made in the twenty-first century in the sense that, even if very locally produced, they could directly confront us with animality as a globally eco-ontological other of the human. Animals in such films, made in Western or Third World countries, appear on the boundary between the Symbolic and the Real, between fictional and documentary aesthetics – that is, at the limit of the symbolic construction of fictive reality. This liminality is another name for animality.

A positivist lessen in this context is found in *Project Nim* (James Marsh 2011), a pure documentary on behavioral scientist Herbert Terrace's 1970s project of raising a chimpanzee called Nim in a human family and teaching it American Sign Language. Not long after the phenomenon of Woodstock, the project took place not in a laboratory but mostly in a huge house with large natural grounds outside New York. The ambitious experiment of bringing human communication to the animal was conducted in the hippie mood of bringing the human back to nature. Its home-movie-style footage indeed shows the most positive communication that could occur "when species meet," if we borrow a book title from Donna Haraway, who draws attention to the everyday practice of intersubjectivity between human and animal and their mutual response in work and play. Notably, Haraway criticizes contemporary Continental philosophy on the animal for ignoring ordinary and mundane interspecies companionship. She argues that Derrida's famous speculation on his cat's gaze at his naked body, which virtually ignited recent animal studies in the humanities, "failed a simple obligation of companion species" by lacking curiosity about "what the cat might actually be doing, feeling, thinking, or perhaps making available to him in looking back at him" (Haraway, *When Species Meet* 20). Gilles Deleuze and Félix Guattari's anti-Oedipal and anti-capitalist project proposed the provocative notion of "becoming-animal," which, however, proves the profound absence of respect for and with actual animals, only "figuring relentless otherness knotted into never fully bounded or fully self-referential entities." Against this "philosophy of the sublime, not the earthly, not the mud" (27–32), Haraway builds on behavior semiotics and ecological biology that shed light on life-entities' *autopoiesis*, the self-making and self-maintaining feedback with other entities in Gaian systems, cybernetic or otherwise.[6] When becoming-animal works as the deterritorialization of human subjectivity into the liberating molecular state immanent in all beings, animal companionship serves the individuation of animal subjectivity in the systemic environment of communication.

Unfortunately, however, Project Nim finally failed despite the deep and long emotional connection made between Nim and his educators. Nim's understanding of sign language was remarkable but limited, while his occasional violence hurt several dedicated animal lovers. Terrace abandoned the project; Nim was sent to a farm and put in a pen with iron bars. The project started on the hopeful fact that 98.7 percent of the DNA in humans and chimpanzees is identical, but its end suggests that the unbridgeable abyss is inherent in the 1.3 percent difference between the two species. Nim's linguistic precariousness and unpredictable violence may all be condensed in this small yet decisive portion of alterity for which humans have nothing to offer but the old alternative: care or cage. That is, care-taking companionship between species could not be as symmetrically mutual as Haraway desires; it would still face the risk of treating the animal other as an object of hospitality or hostility, tolerance or intolerance. Haraway, of course, emphasizes the training in the contact zone for practical interactivities with animals that can challenge human exceptionalism. A fundamental question is, however, who initiates such interactivities. It is always the human and not the animal that desires to bridge their gap, and thereby the human always takes the position of a host who invites animal guests to his home.

The ideal of mutual companionship, then, evokes the idea of impossible hospitality. For Derrida, pure hospitality is impossible in both ways: first, the host cannot be hospitable towards the guests who take over his property ownership and control of the situation, that is, who threaten the precondition of hospitality itself; second, if unconditional hospitality is offered through non-mastery and the abandon of all property, there is no longer the possibility of hosting anyone as there is no ownership or control (Derrida and Dufourmantelle 135–55). Conversely, hospitality is possible only as

conditioned, limited, just like the religious notion of tolerance taking the form of a Christian charity – its paternalistic gesture of invitation still implies the juridical subordination or assimilation of the other to the host's symbolic order: "I invite you, I welcome you into *my home*, on the condition that you adapt to the laws and norms of my territory, according to my language, tradition, memory, and so on." Pure hospitality, on the contrary, is open "to someone who is neither expected nor invited, to whomever arrives as an absolutely foreign *visitor*, as a new arrival, nonidentifiable and unforeseeable, in short, wholly other" (Habermas and Derrida 162). The first case of partial hospitality resonates with Žižek's critical view of capitalist multiculturalism whose liberalist tolerance is limited to benevolent other cultures, music or food, which are deprived of their excessive intolerable otherness: violence, patriarchy, fundamentalism, etc. (Žižek, "Multiculturalism"). The second case of hospitality *itself* implies the complete loss of anthropocentric initiative that sets up the field of interaction as such, returning to pure nature prior to the birth of the notion of nature. If the first is related to the ethical turn of politics in the public sphere, the second insinuates a state in which this civilized *polis* as a cultural frame itself no longer exists. This ideal hospitality is put in a double bind: it is inevitably conditioned in practice through the host's awareness and management of it; or when unconditioned, it couldn't maintain itself as hospitality.

Project Nim tests the limit of such hospitality as cannot but be imperfect in reality. Its concrete trans-species companionship is done through the invitation of the animal to a linguistic community owned and controlled by the human host. The guest who hurts this hospitality is deported and imprisoned, which implies that hospitality cannot be unconditional because of the host's self-protection. In his *polis*, the host can declare a "state of exception" in which the detention of the threatening other is executed in the way of degrading it from a community member to just a "naked life" whose political subjectivity is suspended like the *homo sacer* (Agamben). Politics is underlain by this biopolitics that the host operates through his sovereignty to distinguish the political from the natural state of subjectivity. Yet this bio-*polis* does not mean that the state of exception restores pure nature, but rather creates a sort of simulated natural state where life becomes vulnerably animal and thus its right to live completely depends on the sovereign's power. Nim finally becomes an animal after his long human education, but then he is an encaged animal, a bestial terrorist under surveillance and punishment.

One may be tempted to save such an animal *homo sacer* in the name of its "rights" drawn from the liberal justice tradition. Animal activists, protectionists, anti-fur protesters, and even vegetarians more or less assume animal rights to be legally endorsed and accepted, just like the human rights of the unrepresented rabble, voiceless people, Guantanamo prisoners, and so on. Paradoxically, this kind of political struggle results in the ethical turn of politics; the more social subalterns register as equal subjects in a community, the more inclusive the concept of rights becomes – it becomes no longer a political goal to achieve but an ethical bottom to accept, almost like a Kantian categorical imperative: we ought to embrace suffering others! Cary Wolfe, however, points out that ethical standing and civic inclusion in the "rights" conversation are predicated upon rationality, autonomy, and agency as intentionality of a member of what Kant called "the community of reasonable beings" (Wolfe 127). Ron Broglio argues that "rights" thus presumes community as founded on humanist ideals, which would inevitably set juridical limits to the nature and exercise of animal rights (Broglio and Young). As a result, while the ethical turn expands the established *polis* bio-ontologically, the humanely endowed rights could paradoxically keep animals more or less segregated, degraded, or at best specially treated through "affirmative actions" taken by the enlarged community. Then hospitality would be reduced back to tolerance, the falsely neutral and potentially hegemonic ideal in the Enlightenment tradition of endorsing others.

One should ask here if this humanitarian ideology is not the unavoidable compromise of ethics or the way it is actualized in reality; an ethics that is defined primarily as one's obligation to the other's suffering, and furthermore the human's compassion on the animal's vulnerability. Judith Butler finds humanity in such an empathy-laden ethics towards the neighbor living a precarious life, its vulnerable face we must not kill in the context of post-9/11 biopolitics (Butler xvii–xviii). But this Levinasian ethics as the reverse of narcissistic aggression is grounded on the cultural (thus still anthropocentric) tradition in which the place of the other traces back to that of infinity opened by God's calling to which Christian sacrifice could respond fully. What counts is one's ability to respond to this infinite otherness, one's "responsibility" for the Other.

Werner Herzog's *Grizzly Man* (2005) leaves room for thinking about a different ethics. To begin with, this auteuristic documentary confronts us with the maximum paradox of hospitality embodied in the life of Timothy Treadwell, a self-claimed animal keeper who was, however, devoured by the very grizzly bears he loved for eight summers in Alaska. The point is not simply that his ethical act led to a perplexing horrible end. The film shows a substantial amount of the video footage made and left by Treadwell himself, in which we actually notice some distance between big "sublime" grizzlies in the background (nature) and the grizzly man with a movie camera as well as a small "beautiful" pet in the foreground (civilization). This physical gap is filled only with Treadwell the speaking subject's ceaseless words that set his psychological connection to the bears. It is as though language, the lack of which defines the animal for many philosophers from Descartes to Lacan, built an invisible barrier between two species, disguised as their companionate communication. Treadwell's one-sided verbal love letters to peaceful animals are virtually like a symbolic wall that protects the animal protector from the animal Real, its dangerous unpredictable violence.[7] Despite his proclamation of animal protection, or rather, because of its linguistic humanism itself, the grizzly man thus remains an ecological multiculturalist whose hospitality partakes of patronizing distanciation from the animal other. All his talk to bears might imply this hidden message: "I love you, but please stay there!" Liberalist tolerance works only insofar as others are not harmful. Treadwell's tragedy is that this distance was too shortened to keep him in safety at a certain point, unawares to him. That is, his hospitality could be maintained only on its own self-contradictory condition that it couldn't be unconditional, though it pretended to be.

Interestingly, however, we realize here that what renders "ecological hospitality" impossible is not the human host's abandonment of his ownership and control (Treadwell kept this condition) but primarily the animal guest's complete indifference to the host. No grizzly bear in fact has recognized his hospitality as hospitality; no animal indeed has the concept of and respect for the hospitable human's property and protection. What underlies this animality is *antiphusis* as aforementioned, with nature as dark, violent, rotten, hostile, which appears negative in the anthropocentric frame, but which fundamentally implies neutrality for the human. A skeleton of a bear devoured by another bear and decaying animal corpses that Treadwell encounters prove not so much a certain animal tragedy as the natural contingency of anti-nature that resists our symbolic explication. Through Treadwell, Herzog sees this permanent crisis of nature as its own homeostasis which blocks any sentimentalized politics of nature. And despite his well-arranged narration, Herzog's symbolic language is more into the ecstatic truth of this enigmatic nature through ephiphanic images of animals and thus much less rationalist than Lacan's formulations of *antiphusis* as barring the Symbolic, that is, as still conceptualized in relation to culture (Noys 49).[8] In passing, Wolfe argues that Žižek treats the animal as a mere metonymy for the Lacanian Real and thus his approach is also still anthropocentric without thinking the "distribution" of subjectivity across species lines (Wolfe 125). But what kind of subjectivity could be distributed to the animal if not

vulnerable precariousness? I will go into this issue below, but at this point it seems more important to focus on the total absence of animal subjectivity connected to the human. A remarkable moment in *Grizzly Man* is when Treadwell finds the steaming lump of a grizzly's feces and calls it "Wendy's poop," as if it were a "gift" by which he can feel the bear's inside, that is, the inside of what is outside him. The dirty material trace of anti-natural rottenness is named and appreciated by the man who thereby feels belongingness to the other. This means not an intersubjective gifting as give-and-take economy but a paradoxical revelation that what is given to him as a gift is never given and even acknowledged as a gift by the animal giver, thus never to be returned. Is it not a miraculous example of pure gift irreducible to exchange which Derrida views as impossible, just like pure hospitality irreducible to tolerance (Derrida, *Given Time* 34–70)?

In other words, pure hospitality that is always impossible when offered by the human to the animal might be possible when we rethink it the other way around. The subject of gifting, of hospitality, is not Treadwell but an originally anonymous bear which, however, has no subjectivity related or intended to him, that is, which has subjectivity, if any, absolutely indifferent to him. The animal offers Treadwell unconditional hospitality neither with call, contract, control, nor with property, protection, or precondition. In fact, it is not that the grizzly man invites grizzlies to his home but that he is accepted and nurtured in their home (called a national park), which he visits without invitation. Genuine hospitality is, then, that which can be only recognized, retroactively, by the visitor and not the inviter, in an exceptional state where there is actually no host/guest power structure. Derrida also suggests "a hospitality of visitation rather than invitation," adding that the visit might actually be very dangerous, but "a hospitality without risk, a hospitality backed by certain assurances, a hospitality protected by an immune system against the wholly other" could not be true hospitality (Habermas and Derrida 162). This true hospitality is again almost impossible to realize whereas its significance may lie in that it serves as the conceptual ideal of actual tolerance, if not perfect, still needed in reality – for Derrida, *le don absolu* is also like the impossible ground on which actual exchange economy is enabled just as ungraspable *différance* catalyzes any system of concrete differences. Yet we can go further than this conceptual justification of pure hospitality or gift on the human's side if we posit the visitor not as the host but as the guest of animal hospitality in a zone of indeterminacy between subject and object. Visitation would thus be viewed as an ethical adventure of abandoning one's subjectivity as a host, becoming a volunteer *homo sacer* who can be killed without being sacrificed in anti-nature, and finding oneself to be in an unprepared and unexpected hospitality without any symmetrical exchange or companionship with the other. Does this not suggest an ethics that is not responsible for the other as a vulnerable sufferer but responsive to the other as a pure gift? A truly ethical act might be to accept the other's being in itself as a gift to me in the realization that it is I who is vulnerable and thus virtually accepted by the very other, gifted its unintended hospitality.

Brilliant in this regard is *Tropical Malady* (2004) by Thai director Apichatpong Weerasethakul, whose locality depicted in a surreal as well as hyperrealist (often documentary-looking) style has opened a new territory of world cinema. The film consists of the first half showing a shy gay couple Tong and Keng's euphoric meanderings mostly in a city and the second half unfolding in the jungle into which Tong suddenly disappears and Keng jumps to trace his lost love. This abrupt spatial shift is also temporal, marked by the audacious intermission of a ten-minute blank screen that looks as though "time is out of joint." It ruptures the present, while opening its subsisting past in itself, the mythical memory of the world retained in the jungle. There, it is narrated that a folkloric shaman has transformed himself into a tiger and is terrorizing the countryside that Keng's army protects. In effect, Tong, appearing in the jungle as a naked man who can shape-shift into an animal, must be the very shaman-tiger that undoubtedly devoured him. A baboon tells Keng in an animal language (subtitled) to kill the lonesome tiger to free it from its world, or to let it devour him to enter its world. But all this weird setting does not imply the mere anthropocentric antagonism between animal and human, nature and civilization. Rather, the jungle appears as what Charles Baudelaire called "forests of symbols" which correspond to each other, with a Heideggerian "clearing," an open empty space of the forest where Being is unconcealed. At the end, Keng encounters the tiger in the dark, which has been watching him like the gaze from the Real before his sighting of it (see Figs 1, 2). Perching on a tree, it stares directly forward with its calm, fixed, silent sublimity, evoking the bottomless, unreadable, impassive gaze of Derrida's cat. And just as naked Derrida feels ashamed in front of his naked pet and yet realizes that the binary of naked/clothed itself is humanist, improper to the animal for which the notion of nudity does not exist, so Keng feels first fearful and undressed, but then disarmed and opened to the tiger that does not appear to be simply a bestial enemy to hunt. The animal gaze destabilizes the frame of nature vs. culture and seemingly addresses the man in an unheard inhuman voice, which Derrida compares to God's calling (*The Animal That Therefore I Am* 17–18).

But rather than tracing back to Judeo-Christian theology, we could see here the thickly concealed face of the tiger representing neither a vulnerable other to save nor a hostile other to subdue, but an inert interface to a world larger than human, the Virtual immanent in the Actual in Deleuze's terms, the "plane of immanence" to which Keng whispers: "I give you my spirit, my flesh, and my memories ... Every drop of my blood sings our song, a song of happiness." Of course there is no inter-species dialogue; Keng's giving of himself would rather indicate his acceptance of the animal gaze as an unplanned invitation to nature, his response (without responsibility) to unbounded hospitality from the universe that offers a chance to

Fig. 1. The tiger in the dark, which has been watching Keng. Still from *Tropical Malady* (dir. Apichatpong Weerasethakul 2004).

Fig. 2. The moment at which Keng encounters the tiger. Still from *Tropical Malady* (dir. Apichatpong Weerasethakul 2004).

undress his civilized identity and join an immemorial world. Far from creating a romantic happy ending, this hospitality invites him to a radical dissolution or liberation of subjectivity into a primordial zone where animal and human, body and spirit, matter and memory are all indeterminate in deindividuated happiness. "When species meet," the human now thus tries no longer befriending animals but experiences "becoming-animal" in multiple senses: not only becoming a bare life detached from society (Agamben) and becoming stripped of humanity in front of the naked animal (Derrida) but also becoming desubjectified in Deleuze's terms. Though Keng kneels down and moves like a beast, this gesture may not signal the imitation of an individual organism so much as a life-changing line of flight from the (Oedipalizing) organization of subjectivity towards "the marvelous of a non human life" on the plane of immanence (Deleuze and Guattari 231–34).

Furthermore, as his face loses its identity in the dark jungle, just like the tiger immersed in darkness, becoming-animal involves becoming part of the environment where animality and humanity become indistinct in their molecular state. The tiger is virtually "apparent without appearing," like the "phasmid," that stick insect that Georges Didi-Huberman describes whose body perfectly resembles twigs or leaves so as to incorporate rather than imitate its environment (Didi-Huberman 15–20). And as its etymology shows, the phasmid implies "phantasm" and "apparition" between being and non-being that can in a trice mutate into a dangerous beast, devouring us into the abyss that effaces all ontological boundaries. Likewise, the shaman-tiger, a human-animal that devoured Tong, is about to devour Keng's body or at least eat his soul, lurking like a ghost-shadow in the dark.[9] Now, the ontology of the animal takes on the *hauntology* of the ghost, and the threshold between the human and these ontological others appears all the more fatal because its transgression entails the complete surrender of the master's position. But again, this risk-taking visitation to the matrix of others would be the ecstatic price of dismantling rigidified humanist subjectivity. Keng enters the uncanny realm of immanent connectedness to the animal, the ghost, namely all virtual life, becoming imperceptible and clandestine like them. In short, becoming-animal exercises the ethical act of embracing the animal as gifted. Through its indifferent hospitality man does not so much become an animal as disintegrate into the virtual grounding of all actual beings.

reinforcing subjectivity through becoming-other

In light of this context, one could draw on Herzog whose films abound with animals: a cat witnessing to people like Derrida's pet (*Heart of Glass* 1976), domestic animals resisting human mastery (*Woyzeck* 1979), decayed corpses of horses scattered in the desert (*Fata Morgana* 1971), monkeys besieging a quixotic hero's broken raft (*Aguirre, Wrath of God*

1972), molecule-like rats inundating a vampire's coffin and town (*Nosferatu* 1979) … Herzog's politics of the animal, if any, certainly disturbs anthropocentrism (Sheehan). At this point, however, we should note that, rather than being liberating, these animals often represent the remnants of anti-nature, a dead end of mystical romanticism where the sublime adventure of surpassing humanity fails or faces death. In effect, the encounter with animals does not always and literally lead to radical molecular desubjectification, which is practically infeasible. We therefore need to review Deleuze's anarchic becoming-animal in terms of the more or less actual potential to transform socially organized subjectivity through contact with animals. This contact occurs in liminal space where one's life becomes "bare" in the first place, as in Apichatpong's jungle – his *Uncle Boonmee Who Can Recall His Past Lives* (2010) shows not a beast but a ghost in the jungle, a human phasmid that reincarnates the world's memory of itself. The cave is another significant place where the animal and the ghost reside in the Apichatpong films; one may recall the cave-looking Zone in Andrei Tarkovsky's *Stalker* (1979) too, where those who enter it enter their own unconscious reverie and desire. Slightly differently, Herzog's *Cave of Forgotten Dreams* (2010) deserves attention, a 3D documentary on the Chauvet caves of Southern France containing the oldest known pictorial creations of humankind. Painted 32,000 years ago, these first frescos bear evidence of primitives' life surrounded by animals to watch, hunt, fight, tame, admire, play with, and so on. Notable are some of them depicting the transformation of human bodies into, or hybridization with, animals, which seemingly expresses more than species companionship. That is, what are represented in the cave – humankind's first cultural space – are actually people's bare life and their fantasy of animal life, with the boundary of culture and nature blurring. The initial desire for becoming-animal, if we still use this term, is imaginatively figured here, not disfiguring but refiguring humanity and thus prefiguring its reinforcement through animality in the form of a new species like the cyborg. Even if still unrealistic or futuristic, the upgrading reterritorialization of human subjectivity via non-human others has indeed been more palpably imagined and envisioned than its radical deterritorialization since the era of the caveman.[10]

It is therefore possible to take the opposite direction of visitation to nature as deindividuation, a direction in which humans can embody ontological others, now including the machine that takes the place of the animal in technological civilization. This turn may be less an antithetic reversion to anthropocentrism than a dialectic reaction to the environment in which, as Wolfe says, subjectivity could be distributed across species lines and thus reconfigured even into certain, unprecedented species. Let us briefly recollect Peter Greenaway's fake documentary *The Falls* (1980) that presents ninety-two victims of the Violent Unknown Event (VUE), which has caused immortality, disability, and ninety-two new peculiar languages. Among the victims whose surnames begin with the letters FALL are those who really "fall" from buildings and Icarus-style homemade wings, a prosthetic device for becoming-animal. Such symptoms of man's metamorphosis into bird recall the hubristic ambition of flying, suggesting the VUE as a modern Babel myth about the end of civilization (the title also invokes fallout).[11] Despite the apocalyptic scenarios it introduces, however, *The Falls* is not a Stanley Kubrick-style black comedy or a dystopian science fiction. The VUE, the attack of the Real, brings about a somewhat jovial disorder of all human systems and causes new hybrid changes to human bodies. That is, the apocalypse was perhaps the VUE itself, time was already out of joint, and what the film shows is a post-apocalyptic new world that reassembles subjectifying apparatuses by producing diverse languages, changes of sex, identity, and skin, physical deformities, and even a dog's becoming-bird. Such a collective schizophrenia partakes of Deleuze and Guattari's revolutionary disorganization of the entire solidified actuality, but multiple modes and actions of becoming-animal do not destroy the organic form of life. Rather, humanity is mutated in mimetic ways of becoming different organisms, which lets us imagine new potentials of retooling our being, often in conjunction with technology.[12]

The spectrum of this organic change and its implication is, however, wide. If becoming-other in *The Falls* is radical but comical, thus somewhat lightly treated, it is seriously staged as a fictive yet painful process in *Black Swan* (Darren Arnofsky 2010), for instance. The story is simple: Nina is a perfect ballerina for the White Swan with innocence and grace, but Lily is better for the Black Swan with guile and sensuality. They compete to be the heroine of Swan Lake who must play both the roles, while this rivalry expands into a twisted friendship that provokes Nina to discover and explore her dark side of violence and sexuality. Obviously we see a variation of the Jekyll & Hyde motif in an Oedipal/Electra triangle: Nina's unconscious desire repressed under her mother-superego's suffocating control explodes through her uncanny double Lily and her artistic director who is a seductive father figure; yet her inner Black Swan's power becomes so uncontrollable that it finally engulfs her White Swan ego, destroying her body. What attracts us to this perverted family drama with its typical black-and-white Hollywood moral is Nina's extreme desire to imitate the Black Swan, the desire for an artistic ideal that is originally impossible to realize but virtually immanent in her unconscious. The finale visualizes all realistic details in which she embodies the character in the form of becoming-animal; her body transforms into a black swan as if this animal inside her casts off its superficial human skin. But this process turns out to be hallucinatory, as the audience in the diegetic space only sees Nina's perfect performance without bodily change. In other words, the Real of becoming-animal appears only in the form of fantasy, and its discrepancy from reality indicates the unbridgeable fissure of subjectivity in itself. This immanent fissure would open the subject's primordial "body without organs," what Deleuze and Guattari term the molecular matrix of any potential organism prior to being organized, yet what the film shows is not this unrepresentable body but Nina as split into two identifiable bodies, human and animal. The fantasy is then all the more painful as it perfectly mimics the organic form of a single animal, betraying the failure of becoming-animal in reality.

This animal mimesis deserves more examination. It is iconic in that Nina resembles the Black Swan and this animal is not real but fictive, so the cinematic depiction of her bodily transformation is basically a digital version of the caveman's painting of animal–human hybrids; the animal icon is a simulacrum produced by the cutting-edge computer technology. However, it looks indexical as well in that Nina's body not only becomes visually similar to a swan in appearance but also physically incorporates the animal's color, shape, skin – the causal traces of her actual connection to material animality – into the iconic simulation. What matters more than the film's visual representation of the swan is Nina's tactile presentation of it, the bodily performativity of becoming-animal which may be all the more dangerous than Treadwell's adventure because of the lack of distance between Nina's and the swan's bodies. This sensational corporeality makes her becoming-other as risky as wanted, and ultimately more passive than active. Her self-centered desire for the swan turns into (and out to be) the other-centered desire that forces her to become it out of her control; she is swept by monstrosity inside her that is more than she is.

This dialectics of active and passive is salient in David Cronenberg's work, whose films have provocatively incorporated the machine along with the animal into the human body. *Videodrome* (1983) now looks like a precursor to *Black Swan*: on the border between reality and hallucination, Max, a mediaholic man, literally penetrates a woman's vulva-like mouth on TV, but this phallocentric aggression is none other than the submission to the screen-body that seduces and sucks the spectator, replacing visual distance with tactile proximity. In analyzing this film, Steven Shaviro argues that the passion for the image is not far from radical passivity, "a forced, ecstatic abjection, a form of captivation" (Shaviro 49).[13] This process entails the sensation of touch as grounding self-transformation. The hero changes into a "video-activated body," becoming ever more biotechnologically transgressive. The slit on his belly becomes a slot for videocassettes, a link between surface (skin, retina, image screen) and volume (convoluted thick entrails), and a vaginal orifice indicating his feminization, as "interfaces between biology and technology run amok" (142–44). The embodied desire for radical tactility ultimately turns his hand into a techno-fleshed pistol, whose trigger he pulls on himself at the end. That is, he destroys the tactile body that blocks him from completely joining the other on TV and its virtual world of videodrome; like Nina, he is stuck between actual and virtual bodies. One could say that his intensive experience of the other on screen reaches the impossibility of becoming the other within the possibility of becoming a mechanical interface to the videodrome; thus, becoming-other is possible only as becoming-interface, which implies becoming-abject in society. Despite the dystopian mood of the film, this technological bare life ends up forever floating as an immaterial ghost image in the videodrome as if Max were liberated from his gendered, socialized mortality. The final manifesto "Death to videodrome; long live the new flesh!" sounds like this: Death to the interface; long live the techno-body! We now need hauntology of technology.

In short, there is a double ambivalence: the subject's apparent aggressivity towards the screen-body entails his passive transformation into an interface, and this becoming-interface entails both abjection and liberation as well as regression and transgression. This ambivalence concerns bio-ontology at large in Cronenberg's films such as *Scanners* (1981), *The Fly* (1986), and *Dead Ringers* (1988), wherein the human body embodies parasite alterity in forms of other minds or even the non-human. Becoming-other, biological or technological, is nothing other than this becoming-parasite that puts the subject in the double bind between self and other. Radically dispossessed and decentered, subjectivity remains all the more vulnerable and constrained so that its "schizophrenic dislocation" involves "bizarre distortions and topographical transformations of physical, corporeal, and social space" (Shaviro 117). In a broader history, the evolution of this body transformation on screen could be traced this way, following Shaviro's cue: (1) Buster Keaton's body combined with machines represents early capitalist mechanism, practical materialism, subversive Dadaism; (2) Jerry Lewis's body reflects late capitalist simulation, multiple and pulverized mass media images; (3) Cronenberg's body reembodies the late capitalist technology of disembodied information and algorithm by morphing into hybrid, corporeal interfaciality, dismantling old dichotomies of mind/matter, male/female, and human/inhuman to show the unshowable monstrous ambivalence between fascination and disgust. The Cartesian–Hegelian epistemology that interrupts immediate embodiment based on the opposition between active subject and passive object now gives way to a fatal phenomenology of embodiment: dangerous passivity is open to "a Bataillean ecstasy of expenditure, of automutilation and self-abandonment – neither Imaginary plenitude nor Symbolic articulation, but the blinding intoxication of contact with the Real" (54).

The question is then: can there be any positive contact with others that reinforces subjectivity in the actual rather than renouncing it? How can one productively transform oneself through otherness while not being entrapped in schizophrenic fantasy, the failure of becoming-other or impasse of the Real? Going back to Haraway, we should recall her "Cyborg Manifesto" that declares we are all theorized and fabricated hybrids of machine and organism (Haraway, "Cyborg Manifesto" 150). Her usage of the futuristic buzzword "cyborg" is metaphorical in an attempt to challenge naturalism and essentialism so that feminists can break away from Oedipal narratives and Christian origin myths like Genesis. She thus sounds somewhat Deleuzian, but more important appears to be the multiple implication of cyborgism: it means the human–machine hybrid, moreover the fusion of both fiction and lived reality, but first and foremost a cybernetic organism, that is, a(ny) self-making organic life as an autopoietic system of feedback through other organisms in the environment. In this sense an animal, a machine, just like a man, is all more or less cyborgian (cybernetics has combined biology with mechanics); the cyborg in SF is a technologically externalized form of this cybernetic mechanism immanent in the environment and shared by various organisms. Then one's connection to this cybernetic network could be governed in order to build up oneself without being disorganized there. Nina in *Black Swan* could have been upgraded into a more powerful female subject – even as a femme fatale – through her becoming-animal without leading to death; through the use of CGI, her sensual transformation even looks like the birth of a sexy cyborg, evoking the female robot in *Metropolis* (Fritz Lang 1927). Most Hollywood films externalize such embodied hybridity in the simplistic negative form of separate antagonistic species, mainly updating the classical Frankenstein motifs: the human hubris of becoming a creator and the subsequent anxiety over technology. Suffice it to recall the war between the machine and the human in the *Terminator* series; in the place of the machine, the *Planet of the Apes* series positions an upgraded animal species. This type of dystopian global cinema obsessively depicts the dominance of ontological others in the wake of global catastrophes as the inevitable outcome of triumphant (biotechnological) modernization.

By extension, finally, *Avatar* (James Cameron 2009), however, serves as a counterexample that envisions a technological future of nature with an animal cyborg introduced. In fact, this groundbreaking digital 3D science fiction with a clichéd narrative has been a target of postcolonial criticism: on a green planet called Pandora (evoking the mythical West), the blue humanoid Na'vi fight human invaders (just as American Indians resisted gold-rushing white colonizers), led by a white male hero outcast from his own race in order to save the ethnic Na'vi community while getting (if not marrying) a beautiful local princess – Žižek, among others, blames the film's multiculturalist stance for being invertedly racist (Žižek, "Return of the Natives"). However, such criticism repeats the aforementioned hermeneutic frame of reading allegory. What is missing is the film's surface level on which we see not the reincarnation of the apache and cowboys but unprecedented amalgams of human and animal on the one side, and of human and machine on the other. The human and the Na'vi literally become part of mechanical- and animal-vehicle-weapons as more than external tools. The soldiers do not simply operate thirteen-foot-tall robots that they ride by means of buttons or joysticks; rather, their muscular action itself is amplified into the robot's physical movement just as a child plays at being a windmill or train by moving its whole body. That is, they indexically embody the machine by performing what Walter Benjamin calls the "mimetic faculty": not the visual or linguistic imitation that maintains a certain transcendent distance from the object, but the tactile incorporation that eclipses such abstraction (Benjamin). This sensational assimilation of the other is more organic between the Na'vi and the animal, as their bodies are so seamlessly coupled that the animal looks less like a useful prosthetic device than like an empowered human itself. In other words, the animal-vehicle is not the natural opposite of the machine-vehicle but its higher version with regard to the human's embodiment of ontological otherness.

Notably, the army's control center is full of high-tech human–computer interfaces, malleable and multilayered digital screens that operate on the human digit's touch and thus reduce the gap between the image and its viewer, who thus becomes a user, player, and conductor of digital interfaces.[14] This tactile indication of information thereby enhances the embodiment of technological interfaces, resonating with contemporary spectatorship in haptic cinema, installation art, interactive games that rearticulate eye and hand, sight and touch, and vision and body. Computer interfaces have actually been increasingly incorporated into the body in this order: punch card, mouse, touchpad, touch screen, gesture sensing, voice recognition, and forthcoming Brain–Computer Interfaces that one can control through cerebral stimulation, as does Jake, the hero of *Avatar*. Input devices for the cybernetic feedback loop thus tend to come into the body. If material interfaces, including Jake's coffin-like pod to plug in his avatar, are still external to the human body, interfaces are indeed internalized for the Na'vi. Their braided hair functions as an embodied interface: its terminus, or its "digit" if we want, consisting of sensitive tendrils are neural links able to mesh with other Na'vi and other sentient creatures as living interfaces such as flying dragons (which make a one-to-one bond with a Na'vi) and trees of souls (which also have biological USB-like links). The Na'vi's sensori-motor capability is maximized, with no artificial mediation, when connected to this empowering fauna and flora. Pandora is a Bio-Internet on which embodied interfaces upload and download data-memories electrochemically, as the roots of the trees communicate with each other like the synapses between neurons. A "global village" is fully fleshed out in this natural network whose mechanism resonates with the historical shift from modern mechanics to postmodern biotechnology (Rosenfeld).

In this background Jake's shift from the human-machine to human-animal side suggests that animality is not prior or inferior to technology but rather posterior or superior. Pandora reflects not so much the nostalgic past of pure nature as a lost paradise but an idealized future of planetary intelligence with embodied interfaces towards which current human civilization is oriented. The sacred Hometree called the Unobtanium might be not merely a post-oil energy source or primitive shamanistic center but literally and figuratively a power "plant" that generates the network of rhizomatically interconnected interfaces.[15] Jake's adventure would be no other than changing his network to a more evolved one. He lives a biological bare life as a result of his paraplegia, and even becomes socially bare after "betraying his own race," which suspends his political subjectivity as a human soldier. But the Na'vi accept him with hospitality, and he himself becomes a gift to a Na'vi woman and her entire community. Since the Na'vi are not indifferent animals but emotional humans, this gifting in positive reciprocity finally leads to Jake's restoration of political subjectivity through his resurrection as a more upgraded human, a member of the Na'vi community. The "virtual reality" interface to his avatar disappears behind the embodied interface of a real Na'vi that he becomes.

The dilemma is life's total dependence on the network without which it cannot be sustained. Individual subjectivity is guaranteed only through deindividuation into monadic agencies acting on, and reacting to, interfaces. From another perspective, an individual is a parasite to the host network site, even when he or she uses prosthetic interfaces from it. Rather than a war between good nature and evil civilization, *Avatar* might show the process by which this vast host fights, defeats, and repels humans as an army of antagonistic parasites like virulent viruses or terrorists. Here, we see the verso of Pandora even though Eywa, "the earth Mother," is the metaphysical Soul that colors this eco-utopia with harmonious wholeness and spiritual plenitude. Anti-nature within nature now takes the form of global complexity immanent in the very networked system itself. Because of the global connection, even small, local damage to the network affects all its parasite-members like a global catastrophe, just as all animals rush to help the endangered Na'vi, receiving the Jake-avatar's SOS at the end. In other words, this last-minute rescue results less from animals' companionship with the Na'vi than from the nature-network's self-regulation and self-protection to maintain its homeostasis. The double side of networking is that the more connected we are, the more contaminated we may be; the more subjective we are, the more subject to others we can be; and the more holistic the network is, the more holes for terrorists it might have. If it is true that networks create

terrorism (Galloway and Thacker), *Avatar* seems to reflect this reality in the near-future global frame of eco-ethical networking and posthuman bio-informatics. Pandora is not unobtainable, but already immanent on our planet.

zooesis and technesis

We have gone through a series of contemporary films broadly in two directions so far. First, it is more fundamentally ethical to accept (or visit) the animal not as a vulnerable other but as a hospitable gift, and thereby open human subjectivity to liberation in the natural environment. Second, on the contrary, human subjectivity can be reconfigured and reinforced through contact with ontological others, including the machine in the technological environment. Dangerous anti-nature beyond anthropocentrism in the former takes the form of terroristic catastrophe immanent in the networked system of the latter. Ontological otherness resides in such an environment as challenging and transforming our being. In this context, we will need to go through both *zooesis* – by which Una Chaudhuri means the discourse on the animal – and *technesis* – Mark Hansen's term for the discourse on technology – in general in their combination, so as to better map and address the global cinema of ontological others. Inspired by Alice Jardin's critique of *gynesis*, a reduction of woman in the service of phallogocentric theory, Chaudhuri criticizes the way the animal is viewed in *zooesis* and proposes artistic ecology beyond anthropocentrism (Chaudhuri and Enelow). More complexly, Hansen argues that even postmodern *technesis* reduces the radical externality and concrete materiality of technology to a mere material support for instrumentality, textuality, subject constitution, or social organization (Hansen).[16] Like the animal, technology is subversive to traditional thought, yet its exteriority is relativized (i.e., the technology of writing; the actualized form of the Real or Virtual) through the ontogenetic mechanism of *différance*, *becoming*, etc. What Hansen pursues is, then, the absolute exteriority of technology that conditions our "noncognitive and nondiscursive *affective* bodily life" as embodied in Benjamin's practice of mimetic faculty (21–30).[17]

Reserving further discussion for a later chance, let me just point out that *zooesis* and *technesis* will continue to be updated in a dialectic struggle to overcome their own limits. For, simply, we cannot help reframe our ontological others in the symbolic discourse. But we also have the image beside language, especially the cinematic image that never stops the movement of confronting us with the outside of any linguistic frame. It deserves noting here that cinema is in essence the amalgam of animality and technology: first, still images are animated, their mobility underlies the foundational "movement-image" that keeps opening narrative space in Deleuze's term, and its genre format is called animation; second, this motion is produced by the cinematic apparatus, stored in the archive of images, which forges and remains an artificial intelligence of the world. It is no coincidence that the zoopraxicope, a pre-cinematic device for displaying motion pictures, created the illusion of animals running inside rotating glass disks, and its inventor Eadweard Muybridge's photographic work on animal locomotion set the history of the moving image in motion. Since its inception, cinema has indeed functioned as an imaginary zoo for animals reentering vision and a scientific laboratory for futures coming in motion. And as aforementioned, the ontological nature of the cinematic image is ontologically other than the living or dead: animate but insubstantial, visual but intangible, spectacular but unreal. Before there is ghost film, film itself is a ghost.[18]

A final note should be about the human's ontological hybridity as a futuristic but imminent *a-venir* of the inevitable evolution of biotechnics and interface culture. Animality and technology no longer form a naïve dichotomy of nature vs. civilization but connect with each other in ways of making more visible the posthuman condition of life. It unfolds in a cinematic "zone" that now goes global, an illusory *clearing* for *bare life* within the globalized world. This zone thus exists like an eco-ontological *heterotopia*, to use Michel Foucault's term, whose identity is not anchored in our society but exists only in the exceptional state of temporary potential to depoliticize or repoliticize any humanistic politics.

Movie going is a visitation to this zone of ontological others where we are ultimately invited to revisit our own being in the world.

notes

1 The common sense that nature is followed by culture is reversed here, but this reversion does not imply the anthropocentric hierarchy that culture is superior to nature. Recollect Jacques Derrida's deconstruction of this hierarchy: culture is a differed-differing nature just as speech is another writing that is *différée* and *différante*; ultimately, *différance* underlies any conceptual opposition which thereby turns out to be the "theoretical fiction" (Derrida, *Marges de la philosophie* 18–25). Our view of reality including nature and culture is itself a constructed fiction, an ideological fantasy whose anthropocentrism is hardly recognized, yet always immanent in our daily life.

2 As Una Chaudhuri and Marina Zurkow say, beloved animals in our pet culture are coddled and pampered at home, shown off and admired in the street, invited to intimate places, given catchy names and special diets, and thus normalized in civic and domestic space. By contrast, when not belonging with humans, animals are made to disappear, are eradicated, excluded, or forgotten.

3 Antonia Levi examines "the werewolf in the crested Kimono" in comparison with its Western counterpart, looking at other Japanese anime/ manga works too: *Phoenix: the Sun* (Tezuka Osamu 1986), *Wolf's Rain* (Nobumoto Keiko 2005), *InuYasha* (Takahashi Rumiko 2004), etc.

4 Adrian Johnston consults Lacan's several seminars of the 1970s, which must have influenced Žižek's idea on nature even when he does not refer to Lacan (Žižek, "Nature and its Discontents"). Below, we will ask whether the Lacan– Žižek line doesn't still conceptualize the Real of nature only in view of the symbolic and thus

cultural frame, but for now I take their schema as a channel to the other side of "our" nature.

5 These interpretations were common in the 1970s–1980s when film theory first and foremost borrowed structuralist semiotics and classical psychoanalysis centering on Freud's Oedipal triangle and early Lacan's mirror stage (Rose; Bergstrom; Bellour).

6 She draws on a broad sense of American pragmatist tradition, including such scholars and scientists as Gregory Bateson, Jane Goodall, Marc Bekoff, Babara Smuts, and Lynn Margulis. Obviously, animal studies in this theoretical background repeats and updates the old dichotomy of Continental and analytic philosophy.

7 In a full review of the film, I pointed out this unnoticed paradox which also resonates with Herzog's ambivalent attitude to Treadwell, attraction and distanciation (Jeong and Andrew).

8 At the end of his essay on Herzog's *antiphusis*, Benjamin Noys expresses a certain anxiety about the dehistoricization of this nature into mysticism, and argues: "To begin to restore a politics of nature involves the restoration of a signifier, and what Herzog provides are the images that call for the re-inscription of the signifier of nature" (Noys 50). But this need for the Symbolic would be only regressively abstract unless it rather goes beyond the very political frame of reducing the impenetrability of anti-nature to simply ahistorical mysticism.

9 An interesting literary reference: at the end of Henry James's short story "The Beast in the Jungle," an uncanny bestial face bursts out of such darkness towards the hero as if it figures his hidden past, his unconscious memory. If the animal is a physical other of the human body, the ghost is a psychological other of the human spirit like one's repressed double. The animal appears in space and nature; the ghost returns through time and memory.

10 Notably, Herzog's 3D image that maximizes the "indexical" nature of the cinema – the image as the physical proof of real objects – looks both excessive and insufficient. For what it captures in enhanced spatial illusion is the cave's 2D wall with painted "icons" of fictive animal-humans, and not their actual 3D entity. A similar dimensional shift from 3D indexicality to 2D iconicity (3D in

the sense of normal cinematic illusory space without using stereoscopic technology) is also found in *Tropical Malady*, just after Keng encounters the tiger: a low-angle panning of the 3D actual forest connects with a horizontal tracking shot of the 2D painted forest depicting the legend of a tiger stretching out its long tongue to a praying man as if to try to devour him. The point is that becoming-animal is in any event still virtual, only iconically imagined, while always inspiring the human (to create "virtual reality" in which it is possible).

11 Such eschatological concerns provoke nonsensical conspiracy theories on the event. For example, one of the victims "Obsian Fallicut had a theory that the VUE was an expensive elaborate hoax perpetuated by A.J. Hitchcock to give some credibility to the unsettling and unsatisfactory ending of his film, The Birds."

12 I somewhere else took an emphatically Deleuzian perspective on this film (Jeong 183–84), but now slightly modify the view. It must be noted in passing that Greenaway has also created a cinematic zoo with diverse animals screened, as seen in *A Zed & Two Noughts* (1986; which means "zoo"), among others, though we do not have space for their analyses here.

13 Vivian Sobchack also discusses the double meaning of passion, "passive suffering" and "active devotion," embodied by both Jesus Christ and *Videodrome*. On the level of prereflective and passive material, "passive suffering" engages us with "response-ability," and "active devotion" with "sense-ability"; on the level of reflective and active consciousness, these correspond to the ethical and aesthetical concepts of "responsibility" and "sensibility" (Sobchack 288–90). Returning to our reformulation of ethics, we could say that the ethics of responsibility for the vulnerable other is less fundamental than the ethics of responsiveness to the hospitable other.

14 Not only such cutting-edge interfaces but also characters' operation of them are a cinematic spectacle and an attractive show in recent SF films including *Minority Report* (Steven Spielberg 2002) *Déjà vu* (Tony Scot 2006), and *Iron Man 2* (Jon Favreau 2010).

15 Ken Hillis points out that the green network of global Pandora has been envisioned through such models as World Brain (H.G. Wells), electronic

noosphere (Teilhard de Chardin), Hive Mind (Kevin Kelly), electronic hyperbody (Pierre Lévy), etc. Why not Google?

16 That is: instrumentality (Heidegger's *Zuhandenheit* (handiness) concerns the usefulness of the tool), textuality (Derrida's *différance* operates the text as machine), subject constitution (Lacan's *objet a* often appears in mass media as technological effect), and social organization (Deleuze's *agencement* means the assemblage of a social machine).

17 This critique seems arguable. Hansen (over) interprets the "text" and "mass media" as the clue to the "machine reduction of technology," though his quotes from Derrida and Lacan do not even contain the word "technology." He also reduces any ontogenesis to his model of technesis, when one may ask how his notion of material technology could be produced if not mechanically.

18 Akira Lippit points out that psychoanalysis, X-rays, and cinema all emerged in 1895, opening the interiority of the mind, the body, and the world respectively. His term "avisuality" revolves around Derrida's idea of spectrality of the image (Lippit; Derrida and Stiegler).

bibliography

Agamben, Giorgio. *Homo Sacer: Sovereign Power and Bare Life*. Trans. Daniel Heller-Roazen. Stanford: Stanford UP, 1998. Print.

Bellour, Raymond. "System of a Fragment (on *The Birds*)." *The Analysis of Film*. Bloomington: Indiana UP, 2000. 28–68. Print.

Benjamin, Walter. "On the Mimetic Faculty." *Reflections: Essays, Aphorisms, Autobiographical Writing*. Ed. Peter Demetz. Trans. Edmund Jephcott. New York: Schocken, 1978. 333–36. Print.

Bergstrom, Janet. "Enunciation and Sexual Difference." *Feminism and Film Theory*. Ed. Constance Penley. New York; London: Routledge; BFI, 1988. 159–85. Print.

Broglio, Ron, and Frederick Young. "The Coming Non-human Community: A Conversation." *Art and Research* 4.1 (2011): n. pag. Web. 5 Nov. 2011.

Butler, Judith. *Precarious Life: The Powers of Mourning and Violence*. London and New York: Verso, 2004. Print.

Chaudhuri, Una, and Shonni Enelow. "Animalizing Performance, Becoming-Theatre: Inside Zooesis with the Animal Project at NYU." *Theatre Topics* 16.1 (2006): 1–17. Web. 30 Dec. 2011.

Chaudhuri, Una, and Marina Zurkow. "Zoöpolis." Web. 20 Aug. 2010.

Deleuze, Gilles, and Félix Guattari. *A Thousand Plateaus: Capitalism and Schizophrenia*. Trans. Brian Massumi. Minneapolis: U of Minnesota P, 1987. Print.

Derrida, Jacques. *The Animal That Therefore I Am*. Ed. Marie-Louise Mallet. Trans. David Wills. New York: Fordham UP, 2008. Print.

Derrida, Jacques. *Given Time: I. Counterfeit Money*. Chicago: U of Chicago P, 1992. Print.

Derrida, Jacques. *Marges de la philosophie*. Paris: Minuit, 1972. Print.

Derrida, Jacques, and Anne Dufourmantelle. *Of Hospitality*. Trans. Rachel Bowlsby. Stanford: Stanford UP, 2000. Print.

Derrida, Jacques, and Bernard Stiegler. *Echographies of Television: Filmed Interviews*. Trans. Jennifer Bajorek. Cambridge: Polity, 2002. Print.

Didi-Huberman, Georges. *Phasmes: Essais sur l'apparition*. Paris: Minuit, 1998. Print.

Feil, Ken. *Dying for a Laugh: Disaster Movies and the Camp Imagination*. Middletown, CT: Wesleyan UP, 2005. Print.

Galloway, Alexander R., and Eugene Thacker. *The Exploit: A Theory of Networks*. Minneapolis: U of Minnesota P, 2007. Print.

Habermas, Jürgen, and Jacques Derrida. *Philosophy in a Time of Terror: Dialogues with Jürgen Habermas and Jacques Derrida*. Chicago: U of Chicago P, 2003. Print.

Hansen, Mark B.N. *Embodying Technesis: Technology beyond Writing*. Ann Arbor: U of Michigan P, 2000. Print.

Haraway, Donna J. "A Cyborg Manifesto." *The Cybercultures Reader*. Ed. David Bell and Barbara M. Kennedy. London: Routledge, 2000. 291–324. Print.

Haraway, Donna J. *When Species Meet*. Minneapolis: U of Minnesota P, 2008. Print.

Hillis, Ken. "From Capital to Karma: James Cameron's *Avatar*." *Postmodern Culture* 19.3 (2009): n. pag. Web. 5 Jan. 2011.

Jameson, Fredric. *Postmodernism, or, The Cultural Logic of Late Capitalism*. Durham, NC: Duke UP, 1991. Print.

Jeong, Seung-hoon. "Systems on the Verge of Becoming Birds: Peter Greenaway's Early Experimental Films." *New Review of Film and Television Studies* 9.2 (2011): 170–87. Print.

Jeong, Seung-hoon, and Dudley Andrew. "Grizzly Ghost: Herzog, Bazin and the Cinematic Animal." *Screen* 49.1 (2008): 1–12. Print.

Johnston, Adrian. "Ghost of Substance Past: Schelling, Lacan, and the Denaturalization of Nature." *Lacan: The Silent Partners*. Ed. Slavoj Žižek. London and New York: Verso, 2006. 34–55. Print.

Levi, Antonia. "The Werewolf in the Crested Kimono: The Wolf–Human Dynamic in Anime and Manga." *Mechademia* 1 (2006): 145–60. Print.

Lippit, Akira Mizuta. *Atomic Light (Shadow Optics)*. Minneapolis: U of Minnesota P, 2005. Print.

Noys, Benjamin. "Antiphusis: Werner Herzog's Grizzly Man." *Film-Philosophy* 11.3 (2007): 38–51. Print.

Rancière, Jacques. "The Ethical Turn of Aesthetics and Politics." *Aesthetics and its Discontents*. Cambridge: Polity, 2004. 109–32. Print.

Rose, Jacqueline. "Paranoia and the Film System." *Feminism and Film Theory*. Ed. Constance Penley. New York; London: Routledge; BFI, 1988. 141–58. Print.

Rosenfeld, Kimberly N. "'Terminator' to 'Avatar': A Postmodern Shift." *Jump Cut: A Review of Contemporary Media* 52 (2010): n. pag. Web. 5 Jan. 2011.

Shaviro, Steven. *The Cinematic Body*. Minneapolis: U of Minnesota P, 1993. Print.

Sheehan, Paul. "Against the Image: Herzog and the Troubling Politics of the Screen Animal." *SubStance* 37.3 (2008): 117–36. Print.

Sobchack, Vivian. "The Passion of the Material: Toward a Phenomenology of Interobjectivity." *Carnal Thoughts: Embodiment and Moving Image Culture*. Berkeley: U of California P, 2004. 286–318. Print.

Wolfe, Cary. *What is Posthumanism?* Minneapolis: U of Minnesota P, 2010. Print.

Žižek, Slavoj. *Looking Awry: An Introduction to Jacques Lacan through Popular Culture*. Cambridge, MA: MIT P, 1991. Print.

Žižek, Slavoj. "Multiculturalism, or, The Cultural Logic of Multinational Capitalism." *New Left Review* 1.225 (1997): 28–51. Print.

Žižek, Slavoj. "Nature and its Discontents." *SubStance* 37.3 (2008): 37–72. Print.

Žižek, Slavoj. "Return of the Natives." *New Statesman* 4 Mar. 2010. Web. 28 Jan. 2011.

But knowing is far weaker than necessity.
  Aeschylus

Stupidity is a scar [...] [A]t the point where its impulse is blocked a scar can easily be left behind, a slight callous where the surface is numb. Such scars lead to deformations.
  Theodor Adorno and Max Horkheimer

~ I believe you are very good.
~ You are right, said the monster. But besides being ugly, I have no sense [*point d'esprit*]: I know I am only a beast.
~ One is not a beast, replied Beauty, for believing they have no sense. That is what a fool never knows.
  Jeanne-Marie Leprince de Beaumont

If they ask me whether such a man is not to be reckoned an ass rather than a man, I reply that I do not know.
  Spinoza

bernard stiegler

translated by daniel ross

DOING AND SAYING STUPID THINGS IN THE TWENTIETH CENTURY
bêtise *and animality in deleuze and derrida*

"do we know who we ourselves are?"

In relation to responsibility, baseness, reason and unreason, that is, both madness and stupidity, the twentieth century would, in philosophy, be the century of the "great stupidity" of Heidegger – a *Grosse Dummheit* which has everything to do with the *baseness* of thinking referred to by Deleuze, and which, here, must necessarily be related to *horror* – that which confronts humanity with the shame of being human.

What would *this bêtise* be – this "*Grosse Dummheit*"? It would consist in both *saying* stupid things and *doing* stupid things – for example, in making philosophical speeches in a political context where *saying* was *doing*, and *letting* happen, and even *encouraging* to happen, for example in "The Self-Assertion of the German University," subtitled "Speech given on taking solemn charge of the rectorate of the University of Freiburg, May 27, 1933," where Heidegger (before himself referring to Prometheus while forgetting his brother Epimetheus, *the forgotten forgetful one who does stupid things*), asks: "But do we know *who we ourselves* are, this body of teachers and students of the highest school of the German people?" (Heidegger 5).

There are times when to say is to do, and this is what John Austin called the performative dimension of language. In certain circumstances, saying something does something to those one is addressing, and thereby creates a new situation: one's speech is an action. Jacques Derrida long meditated on this philosophico-linguistic discovery by Austin, and we shall see that he relates the profession of professing proper to the university professor to this performativity.

I would like in the first place merely to point out that in the case of performative utterances, *saying something stupid* amounts to *doing something stupid*, and that it is also and especially for this reason that, today, the question of the *responsibility of the university*, or of professors professing their profession (to speak like Derrida in "The University without Condition") through more or less performative statements, authorizing themselves from their autonomy and their self-assertive sovereignty, arises *as never before*. As never *before*: that is, before knowledge began to move from "Promethean" technics towards becoming *technology*, itself becoming the weapon of a global economic war that is ruining the planet, and which leads reason to self-destruction via the "torrent of events" that Karl Polanyi refers to in his book *The Great Transformation*.

prostitution of theory, reification and proletarianization

If regression (*Rückschritt*, to step back) is induced *by reason itself* when it becomes rationalization (including that of mass death), leading to "the tireless self-destruction of the *Aufklärung*" (Adorno and Horkheimer xiv), then this self-destruction (*Selbstzertörung*) rests for Adorno and Horkheimer on a *prostitution of theory* that denatures it and sends it into decadence:

> In the operations of modern science, the major discoveries are paid for with an increasing decline of theoretical work, formation and education. (xiv; translation modified)

> The eighteenth-century philosophy which [...] put the fear of death into infamy, joined forces with it under Bonaparte [...] Such

metamorphoses of critique into affirmation do not leave theoretical content untouched: its truth evaporates [...] the official spokesmen, who have other concerns, are liquidating the theory to which they owe their place in the sun before it has time to prostitute itself completely. (xv)

Progress (the *Aufklärung* understood as *progress* of reason) *thus* inverts its sign (*through* this prostitution and as rationalization): "progress is reverting to regression" (xviii).

The *Aufklärung* has failed and requires a leap, a jumpstart, because it has given up developing the *theoretical understanding* of its *reversing* destiny, according to Adorno and Horkheimer:

> By leaving *reflection* of the *destructive side of progress* to its enemies [...] the mysterious willingness of the technologically educated masses to fall under the spell of any despotism, in its self-destructive affinity to nationalist paranoia, in all this uncomprehended senselessness the *weakness of contemporary theoretical understanding* is evident. (xvi; translation modified; my emphasis)

This theoretical weakness was present in 1947[1] – but it seems in 2011 to be *even more present*, and seems to be *more present than ever in the eyes of the younger generations, and not just to the younger generation of philosophers* trained in France at the Ecole normale supérieure or in universities. This theoretical weakness, which also seems present in "all classes of the population" (Marx and Engels 88), all afflicted by what I have analysed as *systemic* stupidity (Stiegler, *What Makes Life Worth Living* 22, 131), has *historically* emerged from the *prostitution* of the *Aufklärung*.

This *prostitution* of reason and theory consists in making them serve rationalization, not only as the secularization of society (in the Weberian sense) but as legitimation, that is, as rationalization in the sense of Ernest Jones and Sigmund Freud. And this inversion of sign, through which reason leads to unreason, and progress to regression, is *justified* under the mask of reason itself, rationalization consisting in posing and in having accepted as a *conclusion* that "nothing can be done," that is, that *there is no alternative*.

This prostitution proceeds, moreover, from a vast *subjugation of individuals to apparatus*, which induces a *regression to minority* affecting "all classes of the population," deriving from what Adorno and Horkheimer referred to as *reification* (*Verdinglichung*), through which the economy (including today's economy of "cognitive capitalism") *disindividuates* individuals:

> The individual is entirely nullified in face of the economic powers [...] While individuals as such are vanishing before the apparatus they serve, they are provided for by that apparatus and better than ever before. (Adorno and Horkheimer xvii)

What is here called "reification" refers to what I, along with *Ars Industrialis* – counter to the dominant understanding of the discourse of Marx and Engels – have tried to understand as a process of *generalized proletarianization* (on the basis of an interpretation of Marx by Simondon [*Du mode d'existence des objets techniques* 15]), a process that *liquidates all forms of knowledge*, including and especially, today, *theoretical knowledge* (and not only *savoir-faire* and *savoir-vivre*, knowledge of how to do and how to live).

This is the process of grammatization, and of the proletarianization of thinking *and of that understanding which thus escapes reason, that is, which escapes the "kingdom of ends"* (Kant) (and this is essentially what Weber's account of rationalization means). While this process of proletarianization may produce a kind of pragmatic intelligence, *métis*, ingenuity, a shrewdness or a cunning through which everyone seems to have become "cleverer,"[2] it in fact leads to a *generalized stupidity* which, in 1944, comes along with the still very recent advent of the culture industry:

> [The mind or intellect] must perish when it is solidified into a cultural asset and handed out for consumption purposes. The flood of precise information and brand-new amusements make people smarter and more stupid at once. (Adorno and Horkheimer xvii)

Regression thus forms a cocktail of *ingenious stupidities* that derive from *cultural consumerism*.[3] In a general way, however, stupidity or *Dummheit* is a *scar of desire*[4] – of which regression is precisely the return to its primordial stage, which is that of the drives.[5]

The fact that reason can regress and self-destruct, that is, lead to its opposite, which is unreason as stupidity or madness, is not unique to our age: the "tendency toward self-destruction has been inherent in rationality *from the first*" (xix; my emphasis).

Stupidity is *never* foreign to knowledge: knowledge can itself become stupidity par excellence, so to speak. And this is so because knowledge, and in particular theoretical knowledge as passage to the act of reason – or more broadly, noesis – only occurs *intermittently* to a noetic soul *which constantly regresses*, and which, as such, is like Sisyphus, perpetually ascending the slope of its own stupidity, given that, as stated by Simonides and cited by Aristotle in *Metaphysics*, "God alone enjoys this privilege" (982b), that is, the privilege of always *being in actuality*, of *never* being *stupid*, of never going down the path of disindividuation, reification and proletarianization.

This is why not only *can* knowledge make *thought base* but it is *essentially* a matter of its *own* baseness – always threatening, always the threat.

epimetheus and sisyphus – "the most cunning of mortals"

> Stupidity is not error or a tissue of errors. There are imbecile thoughts, imbecile discourses, that are made up entirely of truths; but these truths are base, they are those of a base, heavy and leaden soul. The state of mind dominated by reactive forces, *by right*, expresses *stupidity and, more profoundly, that which it is a symptom of: a base way of thinking.* (Deleuze, *Nietzsche and Philosophy* 105)

One would clearly understand nothing of these lines by Gilles Deleuze extracted from *Nietzsche and Philosophy* if one did not posit, with Dork

Zabunyan, that "stupidity [*bêtise*] must therefore be understood as my own stupidity [*ma propre bêtise*]" (Zabunyan).[6]

This is above all a question of *my* stupidity such that *it* is capable – *that is*, such that *I* am capable – *of making me ashamed*: a stupidity such that *I perceive my being stupid*. Without which (for want of *being* stupid, of being *able* to be) I would not be able to be *affected* (pained, struck) by the *stupidity of others*, or to have shame *for myself* (as if their stupidity necessarily and immediately becomes mine): without that, I could not be *made* ashamed.

It is from out of this *experience of shame* that I begin to philosophize, writes Deleuze in his reading of Nietzsche – and this means that stupidity is "a properly transcendental question: how is stupidity (not error) possible?" (Deleuze, *Difference and Repetition* 151). This is the question of *individuation* and *disindividuation*. If we are *able* to be stupid, it is because individuals only individuate themselves from out of *preindividual funds* from which they can *never break free: from out of which, alone, they can individuate themselves*, but within which they *can also get stuck, bogged down*, that is, dis-individuate themselves.

> [Stupidity] is possible by virtue of the link between thought and individuation [...] Individuation as such, as it operates beneath all forms, is inseparable from a pure ground [or fund] that it brings to the surface and trails with it [...] Stupidity is neither the ground nor the individual, but rather this *relation in which individuation brings the ground to the surface without being able to give it form.* (151–52; my emphasis)

That is, it cannot produce what Simondon called "taking form" (Simondon, *L'Individu et sa genèse physico-biologique* 46). Such a fund or ground may be that of knowledge itself, knowledge that has become "well-known,"[7] and the *best* thoughts – those that *make* knowledge, that open what I describe as a *new circuit of transindividuation*. And yet the *best thinking always remains susceptible to regression*.

The question of stupidity is the question of regression (of lowering, of baseness) in relation to this solemnity [*gravité*] with which thought progresses, that is, raises itself in climbing [*gravissant*] that which is high, in advancing towards what Simondon called "key-points" (*Du mode d'existence des objets techniques* 164) – but always with the risk that inevitably accompanies elevation, the constant imminence of the fall, of which the tightrope walker is the figure (Nietzsche, *Thus Spoke Zarathustra*): the fact that one who thinks only thinks noetically *intermittently* means that the one who thinks, *this one* who thinks, always ends up falling back again, that all thinking can become stupid, eventually becomes stupid (again), and that any knowledge can end up *justifying* and *rationalizing* the *worst* stupidity.[8]

This relation stupidity/knowledge, *bêtise/savoir, Dummheit/Wissenschaft*, such that *knowledge will never be done with stupidity* in so far as it is firstly and *above all* its own stupidity – that stupidity *proper to knowledge*, that is, the *impropriety* of knowledge (that which is taught to us by the figure of Epimetheus and by *epimetheia* which only thinks *from out of* its own stupidity, posing stupidity as its *point of departure*, and which provided the name for the collection *Epiméthée* founded by Jean Hyppolite at Presses Universitaires de France) – this relation, stupidity/knowledge, is what is at stake in what, relying notably on Jacques Derrida and Paul Valéry, I have tried to think as the *pharmacological condition* of knowledge, that is, of *noesis* as the existence possible for non-inhuman beings faced with the fact of being-inhuman (faced with the shameful, and as deficiency of shame, absence of shame, of *aidos*) (see Stiegler, *Taking Care of Youth and the Generations* 180–81).

The pharmacological (that is, Epimethean) condition of knowledge and of *noesis* is equally *that of the university* in so far as it is *an institution in constant struggle against stupidity*, and more particularly against its *own* stupidity (which is always already expropriated, beginning with what Derrida analysed as exappropriation), constantly climbing *anew* the gravity of this pharmacological condition – in order to refound, like a "happy Sisyphus" (Camus 110–11), the meaning and value of the universal which derives its name from *universitas*, that is, *such that, in the universe, something has still not happened, remains still to be climbed* [*à gravir*] *in gravity* ... and to be engraved [*à graver*], according to the mnemotechnical condition described in the "Second Essay" of Nietzsche's *On the Genealogy of Morality.*

derrida plays the fool – and deleuze is not *exactly* derrida

Derrida has commented on this passage in which Deleuze asks how stupidity is possible, firstly in relation to the question of the animal, basing his response on the beginning of Deleuze's argument, where he proposes that "Stupidity [*bêtise*] is not animality. The animal is protected by specific forms which prevent it from being 'stupid' [*bête*]" (Deleuze, *Difference and Repetition* 150).

Following the "well-known" method of deconstruction, Derrida, in *The Beast and the Sovereign* – a work that derived from a seminar that was part of a series dedicated to the question of responsibility – tries to reduce Deleuze's reasoning to a classical opposition between the human and the animal. He thus challenges the possibility of identifying this "property of man" that according to Deleuze stupidity would constitute.

Now, I cannot help but think that Derrida, here, is *playing the fool, fait la bête,* as one says in French, when, for example, he writes that "Deleuze intends to separate man from animality as to *bêtise,* saying without equivocation, decidedly and determinedly, that '*bêtise* is not animality'" (Derrida, *The Beast and the Sovereign, Volume 1* 180). It is hard to understand why, if this is this case, Derrida himself declares, at the beginning of the first session of his seminar devoted to the beast and the sovereign, that "the beast is not exactly the animal" (1).

One could no doubt respond that Derrida reproaches Deleuze by saying something that is *close* to what Deleuze says, but that is not *exactly* what Deleuze says, just as the beast,

according to Derrida, is "not exactly" the animal – just as this odd animal who is Deleuze is *not exactly* Derrida, however much the latter plays the fool [*fait la bête,* acts the beast, plays the fool, makes a blunder]: Deleuze does not say exactly what Derrida says, Derrida tells us, because he says it "without equivocation, decidedly and determinedly."

Now, beyond the fact that *decided and determined clarity,* which is not always useless or harmful, does not necessarily always lead to a logic of *opposition* – it can and even *must* be the clarity of a *distinction* – beyond the fact, also, that the verb "to be" in Deleuze, as in Derrida, is a copula[9] *that we cannot do without* at the very moment when we *want* to deconstruct and to *deconstruct this very impossibility of doing without it,* as, for example, when one says that "*bêtise is not* animality," or that "the beast *is not exactly* the animal" (and it would here be necessary to deconstruct the question of exactitude,[10] and everything that this raises – a thousand necessary tasks which, *equivocating,* could nevertheless end up resembling what Hegel called *Räsonieren,* quibbling), beyond all this, I believe that here, *Derrida totally misinterprets the discourse of Deleuze, and that he profoundly misunderstands the provenance of this discourse on individuation (and disindividuation) in repetition* that is the book *Difference and Repetition.*[11]

Deleuze tried to think stupidity *from out of individuation* – that is, with Simondon. Individuation, he writes, is "inseparable from a pure ground that it brings to the surface and trails with it" (Deleuze, *Difference and Repetition* 152). And it is in relation to this inseparable ground [*fond*] that stupidity takes place as a transcendental structure of thinking.

To develop his argument against this analysis, Derrida focuses on one sentence where Deleuze posits that animals "are in a sense forewarned against this ground, protected by their explicit forms" (152).

But *what is this ground?* Is it, for example, what Derrida called the "groundless ground (*Urgrund* as *Ungrund*)" (Derrida, *The Beast and the Sovereign, Volume 1* 180)? Nothing is less certain. It is a matter of the ground of what Deleuze named *individuation of any kind whatsoever*, that is, whether it be *animal or human*, that "operates beneath all forms, [and which] is inseparable" from such a ground. Animals, according to Deleuze, would, however, be "in a sense forewarned against this ground, protected by their explicit forms."

To this assertion Derrida objects that "if they are forewarned, then they must be in a relation, in some relation, with this ground and the threat of this ground" (180).

Deleuze would not, of course, deny this, and he himself speaks of a relation. But what are these forms that "forewarn" or "forearm" animals *from their ground*, of which we continue to ask (we have not yet had an answer) in what this ground (or fund) consists? And why are they "explicit"?

All *individuals* (humans, animals, vegetables and crystals, that is, minerals) are *individuated* through an *individuation process*. In the vital individuation process, the *genuine individual* is the *animal group that forms the species* in so far as it is affected by that which, in its vital preindividual fund (or ground), constitutes the mark of a phase difference (that molecular theory, for example, relates to copying errors that give rise to singularities, which, within their milieu, may in extreme cases lead to monstrosity or to a mutation of species): the phase shift is not marked in the vital individuation process at the level of the animal, but rather at the level of the animal group that constitutes and individuates the species.

The individuated species constitutes the vital unit, writes Simondon:

> The group is integrative. The only concrete reality is the vital unit, which in certain cases can be reduced to a single being and in other cases corresponds to a highly differentiated group of many beings.

This is particularly visible in termites:

> Thus, termites construct the most complex edifices in the animal kingdom, despite the relative simplicity of their nervous system: they act almost as a unique organism, working as a group [...] What we refer to as the individual in biology is in reality in some ways a sub-individual more than an individual; in biology, it seems that the notion of individuality is applicable at several stages, or at different levels successively included within each other [...] The unity of life lies with the complete group, not the isolated individual. (Simondon, *L'Individuation à la lumière des notions de forme et d'information* 157–58)

In *Civilization and Its Discontents*, Freud highlighted that:

> we should not think ourselves happy in any of these animal States or in any of the roles assigned in them to the individual. (123)

In the language of Simondon taken up by Deleuze, it could be said that we *consider ourselves happy to find material with which to individuate ourselves psychically*.

repetition as individuation

Derrida does not understand the meaning of the words *fond*, *rapport* and *individuation* as they are used in *Difference and Repetition*. That animals are "forewarned against this ground" does not mean for Deleuze that they *are not* in relation to this ground: it means that their relation to this ground passes through *specific organizations*, where the word "specific" means that which characterizes an animal *species*, as *specific relations typical and determinate for this or that animal species*, constituting as such "explicit forms," that is, recognizable forms (including by the animals themselves as *imago* – which makes it possible for the locust to adopt its "gregarious" form, as Lacan says in "The Mirror Stage" [Lacan 77]) and describable forms, *through which the preindividual fund from which they come individuates itself diversely and specifically* – that is, *at the level of the living group* that constitutes a species – and without the isolated animal individual itself being affected by an *indetermination*.[12]

It is here completely *impossible* to follow Deleuze's reasoning without referring *in detail* to the Simondonian philosophy of individuation – which Derrida seems totally to ignore. The "explicit" forms that *species* form (as "taking form") are the *processes of vital individuation* of which the "concrete" forms consist in *processes of specification*. In the first chapter of *Difference and Repetition*, "Difference in Itself," *specification* and *individuation* are linked together by Deleuze, both with reference to Simondon, as when Deleuze paraphrases Simondon by asserting that "the individuating is not the simple individual," and against Duns Scotus, about whom he nevertheless states:

> [Duns Scotus was] not content [despite this] to analyse the elements of an individual but went as far as the conception of individuation as the "ultimate actuality of form." (Deleuze, *Difference and Repetition* 38)

If it is necessary to pass through the thought of Simondon, this is because:

> We must show not only how individuating difference differs in kind from specific difference, but primarily and above all how individuation properly *precedes* matter and form, species and parts, and every other element of the constituted individual. (38)

One of the principal aims of *Difference and Repetition* is precisely to think this link *other than* according to tradition and everything that follows it up until Heidegger (who was also a reader of Duns Scotus) and beyond: it is a matter of thinking *with Simondon beginning with the animal* and, more generally, with the vital – the animal and the vital being themselves thought beginning with the crystal, that is, with the individuation of the mineral. The regimes of individuation are here *kingdoms*, that is, *forms of sovereignty* of which the juridico-social form is therefore merely a case – given that *individuation in general is sovereign*.

In the passages of *Difference and Repetition* commented on by Derrida (which it is hardly possible to read without referring to the passage from *Nietzsche and Philosophy* – published six years before *Difference and*

Repetition – that I cited above, which is also to say, without referring to what Nietzsche wrote about the relation between philosophy and *bêtise* in §328 of *The Gay Science*), Deleuze speaks of a process of *vital individuation* on the ground of which, *from out of* the *funds* of which, and *in* which appears an individuation process *of a new type*: psychic and collective individuation, *which no longer has the same relation to this ground or fund* because it constitutes, precisely, a new *regime* (that is, a new *kingdom*) of individuation.

Individuation in general must be thought as *relation and process* and not as *stasis and identity*. What is new is the relation between the determined and the undetermined, and the way in which they are instantiated in different types of individuation (mineral, vital, psycho-social[13]). The question of the undetermined is posed by Deleuze in *Difference and Repetition* above all in reference to Kant – and to the question of the "I think." In Simondonian thought, this becomes the question of the "phase shift" which constitutes the dynamic principle of the process itself, and which is concretely expressed as the "taking form" of an individuation in an individuated being.

There is a *common ground* to all individuation processes, which are not at all opposed to one another by this thought, contrary to what Derrida would have us believe. But there is a *new relation* to this ground or these funds with each new type of process (mineral, vital, psycho-social), this relation consisting in the *distinction* and the *inscription* of a *difference* – and which is, in addition, a new regime of *différance* – and which itself derives from a *repetition* (and I shall return to this below).

This is why Deleuze can write:

> The animal is protected by specific forms which prevent it from being "stupid" [*bête*]. Formal correspondences between the human face and the heads of animals have often been composed; in other words, correspondences between individual differences peculiar to humans and the specific differences of animals. Such correspondences, however, take no account of stupidity as a

specifically human form of bestiality. (Deleuze, *Difference and Repetition* 150)

Between the human and the animal there is a *change of regime* of individuation which is a *change of relation* to its preindividual funds. Humans individuate *psychically* whereas animals individuate *specifically*.

If "individuation as such, as it operates beneath all forms, is inseparable from a pure ground that it brings to the surface and trails with it" (152), this is because it is always associated with its milieu, which must be understood as a potential for individuation, that is, as a preindividual fund. This potential constitutes *possibilities*, and *it is from these possibilities that it is necessary to think being*, and not the other way around.[14]

indeterminacy and determination – *the wanderer and his shadow* in psycho-social individuation

The preindividual is conceptualized by Simondon through the analysis of crystallization as the individuation of the mineral:[15] the crystal congeals (crystallizes) and stabilizes a tension coming from a metastable milieu that Simondon thinks in terms of the pairs "wave or particle, matter or energy" (Simondon, *L'Individuation psychique et collective* 15), whereas a living thing is an *incomplete and unfinishable* form of mineral. A living thing is a crystal *that does not take*, which is "in between," in a situation of metastability, between stability and instability, engendering a succession of *specific metastable forms* that concretely express this "perpetuated individuation" (16).

This *vital incompleteness* that perpetuates the individuation process instead of congealing it as a crystal establishes and metastabilizes a *situation of différance* – it is this situation which, constantly forming and de-forming, that is, *differentiating itself*, and thus *perpetually individuating itself*, and in *struggling* thus *against its crystallization*, that is, against its pure stabilization, against its hardening, if not its "stupidity" ("stupidity" being a *psychic* and transcendental trait *in that* it is

not a *specific* trait, not the trait of a *species*: the head of the animal is not a good representation), results in the passage from the mineral to the biological (16, and Simondon, *L'Individu et sa genèse physico-biologique* 115ff.).[16]

Since "stupidity" is a transcendental trait, that is (in Deleuze), psychic rather than specific, "[c]owardice, cruelty, baseness and stupidity are not simply corporeal capacities or traits of character or society; they are structures of thought as such" (Deleuze, *Difference and Repetition* 151).

Nevertheless, these structures *of thought* must be thought *from out of the psycho-social* (that is, *both* psychic *and* social) *preindividual ground or funds* such that:

> The individual distinguishes itself from it, but it [this psychic *and* social preindividual fund] does not distinguish itself, continuing rather to be wedded to that which divorces itself from it. It is the indeterminate, but the indeterminate in so far as it continues to embrace determination, as the ground does the shoe. (152)

That the psychic individual cannot psychically individuate itself without socially individuating itself is, in Simondonian theory, the proper trait (specific, this time, in the logical sense of this word) of psychic and collective individuation, but this dual individuation always operates in an intermittent tension between the psychic individual and the social group from which it cannot be separated: from which it can only distinguish and "divorce" itself while remaining "wedded" to it.

It is *also* in this sense that *The Wanderer and His Shadow* must be read, where man

> is *always* living in manifold dependence but regards himself *as free* when, out of long habituation, he *no longer perceives* the weight of his chains. (Nietzsche, *Human, All Too Human* 306)

This *freedom*, however, consists in *forging and adopting new chains*. If animals "are in a sense forewarned against this ground, protected by their explicit forms," this is because they *are not* chained *in this way* – given that it is only possible to *have* chains if it is possible to not *be* chained, and therefore to *make* and *adopt* new ones. Animals, through their species, "are" not a species: they *are* this species.

If man can suffer (from *having* that which he *is not*), then it is "only from *new chains* that he suffers: – 'free will' really means nothing more than not feeling his new chains" (306; translation modified). Now, there are such chains because, from out of the *psycho-social* preindividual funds, psychic individuation *and* collective individuation are *simultaneously* arranged, according to Simondon, and all this presupposes, as I feel it necessary to add at this point, a *technical* individuation. Psychic individuation and social individuation (of the group) can, however, be *turned* against one another, and nullify one another: their confusion is their *mutual disindividuation*, and it is precisely this confusion that leads to *stupidity, bêtise, Dummheit*[17] (baseness), yet psychic individuation and social individuation can never individuate *without* each other, which is what the eighteenth century called *intelligence*.

It is *by separating in a new way that which links them together* (as a new *relation*), thus establishing a *new form of phase difference* in the process of individuation (which is *always* changing phase, since otherwise it would not be a process, that is, a dynamic system rather than a determinist system), it is through this binding separation (the purest form of which is friendship, in the sense given to this by Blanchot[18]) that these psycho-social preindividual funds make possible *a new type of incompleteness* and constitute *through this* a *new regime* of individuation which produces the *transindividual* – that is, meaning.

Even though he did not himself thematize the necessity for technical individuation to support psycho-social individuation, Simondon did propose (in *Du mode d'existence des objets techniques* 248) that the transindividual presupposes artefacts, technical objects, which are also image-objects[19] that must be understood as hypomnesic supports, *hypomnemata, pharmaka* and everything that Derrida analysed as *supplementarity* in that *history of the supplement* that Derrida did not himself ever actually carry out. Derrida did not carry out this history of the supplement even though he announced it, a history which, in the language of Simondon, pursues individuation – in what we should perhaps call the *psycho-social kingdom* – by compensating for an incompleteness that is *other* than that of the living, even though that is its pro-venance, just as the vital has *its* pro-venance in the mineral.

différance and repetition

Simondonian thought *overcomes* the oppositions between types of individuation by referring to traits common to all individuation processes – always constituted through the *individual/super-saturated milieu* (crystalline, vital, psycho-social) which *exceeds the opposition inside/outside* – and such that *these traits individuate themselves* through *types* of individuation and as *relations* in what Simondon called an "ontogenesis," but which I, however, prefer to call a genealogy: the genealogy of different *regimes* of individuation (different kingdoms which are forms of sovereignty – including within species, including within that species called human, or rather non-inhuman within inhuman being, including between psychic individuals, and so on) as local individuations within a much broader process binding and connecting them together.

Such a thought *of individuation as process* is not foreign to that *process that différance also is* – this "kind of gross spelling mistake"[20] on the basis of which alone *it is possible and necessary, in the eyes of Derrida* (I was convinced of this from the moment I began to read it), *to think "gross stupidity,"* that gross stupidity through which was expressed, historico-politically, that hyper-metaphysical sludge in which citizen Heidegger got bogged down (and disindividuated himself, that is, *betrayed himself,* in *both* senses of the term). This relation between individuation and *différance* is something of which we can easily be convinced if we re-read, for example, the following lines:

> *Différer* [...] is to temporize, to take recourse, consciously or unconsciously, in

the temporal and temporizing mediation of a detour that suspends the accomplishment or fulfillment of "desire." (Derrida, *Margins of Philosophy* 8)

"To differ" is in this sense, which is that of *différance*, to implement the *structural incompleteness* of the vital or psycho-social (but not mineral) individuation process such as it was thought by Simondon: "this temporization is also temporalization and spacing, the becoming-time of space and the becoming-space of time" (8). In other words, this *individuation* that is *différance* gives *a difference which spatially concretizes this différance* "to be not identical, to be other, discernible, etc." (8).

The individuation of differences through *différance* is only possible through an originary *phase difference* which is also a default of origin that spaces itself (out) by repeating itself (from out of a *primordial repetition*[21]):

When dealing with *differen(ts)(ds)*, a word that can be written with a final *ts* or a final *ds*, as you will, whether it is a question of dissimilar otherness or of allergic and polemical otherness, an interval, a distance, *spacing*, must be produced between the other elements, and be produced with a certain perseverance in repetition. (8)

problematization of the living

Within the broader processes of individuation, *regressions* are always possible. This does not mean that psycho-social individuation could devolve to vital individuation, that is, specific individuation, or that vital individuation could devolve to mineral individuation.[22] It means, rather, that *psycho-social* life oscillates between dynamic possibilities which characterize types of individuation without separating them:

The psychic and the vital cannot be distinguished like two substances, nor even as two parallel or superimposed functions; the psychic acts as a brake, decreasing the speed of the individuation of the living, a neotenic amplification of the first state of

this genesis. (Simondon, *L'Individu et sa genèse physico-biologique* 163)

As is now well known, neoteny, in the theories of Kapp then of Canguilhem and Leroi-Gourhan, is thought as "organic projection" and "process of exteriorization," that is, as technicization of the living and "technical form of life." Neoteny does not only mean that the living thing needs artefacts in order to live – which is already the case for certain living things which modify their vital milieu by imprinting their form of life upon it. It means that, "if the living being could be *completely peaceful and satisfied in itself*" (163; my emphasis), as "the animal does not reason or work" (Bataille 121), and is in this sense *sovereign* – in a sense which is not that of the psycho-social kingdom where sovereignty derives on the contrary from a primordial inquietude and dis-satisfaction – "there would be no appeal to the psyche" (Simondon, *L'Individu et sa genèse physico-biologique* 163).

Psychic and collective individuation is what occurs when "life problematizes itself" (163). This problematization results in a *decoupling between perception and action*, that is, it means behaving *differently*, *otherwise* than merely a reaction, becoming through that an *act*, an action as *passage* to the act. And this constitutes a transformation of affectivity: affectivity itself becomes *emotion* as the *différance* of the effect from the affect, a *différance* which retains (a retention) and reflects, which psychically individuates – but in transindividuating as the work of the *psycho-social regime of différance*.

This is why there are not

beings that merely live, and others that are living-and-thinking: animals probably occasionally find themselves in a psychic situation. Such situations which lead to acts of thinking are, however, less frequent in animals.

In man, on the contrary, it is

the purely vital situation which is [...] rare [...] There is no nature, no essence on which to found an anthropology; simply, a threshold is crossed: animals are better equipped to live than to think, whereas man is better equipped to think than to live. (163 n. 6)

Dissatisfaction is a *new modality of incompleteness* (of *différance*) through which the living individual becomes a psychic and social individual. Psychic *différance* is immediately social individuation because "the vital functions cannot solve the problems posed by living" (164).

To the extent that he posits explicitly and in principle that technical concretization is the condition of appearance of the transindividual, it is very surprising that Simondon does not engage with the process of exteriorization theorized by Leroi-Gourhan as the consequence of neotenization, that is, of the *technical problematic* – of technical problematization – in the *différance* of psychic life.[23] For the psychic individual only individuates when the resolution of the problems of life, having become *psychic problems* – because the *neotenic living thing* that is the *psychic living thing* can no longer solve them – can only be concretized through *participation in the transindividual* which the psycho-social constitutes and which itself presupposes technical objects that, as object-images, are the supports of the transindividual.

The transindividual occurs to the strict extent that "entering the path of psychic individuation requires the individuated being to surpass itself" (164), and this surpassing of the psychic individual is not only a trans-formation of the self. The self can only trans-form itself psychically to the extent that it trans-forms its social milieu. In order that its psychic transformation can *in fact* become its own, it must transindividually surpass itself as social transformation, that is, as social *différance*:

> The psychic results in a transindividual order of reality [...], the psychic is born of the transindividual [...], psychic reality is not closed in on itself [...], the resolution of the intra-individual psychic problematic [...] occurs at the transindividual level. (164)

That the psychic individual may, however, get bogged down in the transindividual, and therefore function as a *quasi-specific individuation*,

is not only something that *can* happen to the psychic individual: it is a *condition* of its psychic individuation to the extent that it must become collective individuation, and in this *necessity* lies the "transcendental" character of stupidity.

This does not mean that stupidity would be a fall of the psychic individual into a disindividuation which would be the passage to the social – as is the case in Heidegger with the "falling prey" (*Verfallen*) of *Dasein* – since, as we have seen, *this passage to the social* is on the *contrary*, as *collective* individuation, the *condition* of psychic individuation. This means, however, that participation in the transindividual *can in fact fall* into an *interindividuality* within which individuation is suspended:

> Interindividuality is an exchange between individuated realities who remain at their level of individuation, and who seek in other individuals an image of their own existence parallel to this existence. (165)

It is from such a *degradation of the transindividual into interindividuality* that psycho-social individuation can *regress* to a stage which is *neither animal, nor vegetable, nor mineral*. This regression of psycho-social individuation constitutes a *deficient relation to the potential that its preindividual funds constitute* (at once as crystalline, vital and psycho-social: the psychic individual which disindividuates suffers psychically, and somatically, which means that he or she also tends to disindividuate vitally, that his or her organs are in contradiction, and even that they may no longer metabolize, that is, assimilate minerals and so on – this being the preindividual potential for vital individuation).

three types of psychic disindividuation

In psycho-social individuation the specific group gives way to what Freud described as the horde, then, *in constant and functional relation to the prehistoric*, then *proto-historical and finally historical evolution of the hypomnesic supplement*, to what Leroi-Gourhan referred to as the socio-ethnic group, which itself gives way to the socio-political group: psycho-social individuation is characterized by the fact that *it constantly techno-logically modifies the conditions of its individuation* – that is, of its *trans*individuation.

But these successive stages are always tending to return to vital forms of individuation, which constantly polarizes them – they are always tending to put in place the regime of the *specific group*, and to *operate the technical envelope of this group as an animal society* in which the psychic and the social de-compose (and disindividuate) in being superimposed in an interindividuality of the group, which thus becomes more like a herd. This does not mean that technicity is regressive. It means that it constitutes a polarity *at once* regressive and progressive.

Simondon, therefore, says the following about the collective formed by psychic and collective individuation:

[it is a] transindividual reality obtained through the individuation of preindividual realities associated with a plurality of living things [becoming through that psychic individuals], distinguished from the *purely social* and from the *purely interindividual*; the purely social exists, in fact, in animal societies; it is not necessary for a new individuation to exist to expand vital individuation; it expresses how living things exist socially; it is vital unity at the first degree which is directly social. (*L'Individu et sa genèse physico-biologique* 165; my emphasis)

In its *interindividual modality*, and when this spreads to the totality of the social group (through some kind of mimetic contagion), the transindividual psycho-social tends thus to rejoin the "purely social" of animal societies in so far as they are conditioned by a specific individuation (that is, herd-like – in the sense that Lacan refers to the *grégarisation* of the locust, the way it is able to take on its gregarious form) rather than a psychic individuation.

Now, *stupidity always passes through this tendency*, in so far as it seeks to stabilize in the form of an *identity* that which is in reality always a *metastability* with the potential for altering. As such, it is also what *conditions* the formation of an *I*, or of an ego, that is, of a narcissistic structure which mirrors (with) other similar structures in interindividuality – the fantasy of identity being thus constructed as a "narcissism of minor differences" (Freud, *Civilization and Its Discontents* 114), and founded on the paralogisms conceived by Kant long before Freudian analyses of the psychic apparatus and psychic functions.

In the epoch of psycho-power and psycho-technologies, and now with neuro-power, marketing exploits these tendencies in order to take control of the processes of transindividuation – thereby provoking massive disindividuation processes. Given that the projection of a phantasmatic identity polarizes the interindividual, and that the interindividual always haunts the transindividual, the *I* and the ego are thus moments of disindividuation. But this does not mean that we ought to try and reduce or dissolve them (that is, raise them into a dialectical synthesis), if only because disindividuation is the condition of a new individuation, which itself consists in the fabrication of "new chains."

It is necessary in fact to distinguish *three types of disindividuation*:

- *that which proceeds from this interindividuality* where the social group regresses to the purely social, through which it again takes on specific traits (in the sense that they characterize that species of vital individuation) which infest the *I* or the ego;
- *that which occurs as a divestiture by technics* – what Simondon described as proletarianization;
- *that which is necessary for individuation* as the *epokhé* of an earlier individuation,[24] and through which the psychic individual accomplishes a "quantum leap," that is, crosses a threshold in their psychic transformation.

This third form of disindividuation, as condition of the pursuit of individuation, itself presupposes *emotion* as the psychic modality and *différance* of affectivity. It is the "capacity of the individuating being to temporarily disindividuate" (165).

I myself argue that, in the final analysis, *these three forms of disindividuation can never be separated*, and always constitute three necessary moments of psychic individuation in so far as it leads to the formation of the transindividual, that is, of the psycho-social collective individual, and in so far as they must be thought in terms of a *doubly epokhal redoubling* over-determined by technical evolutions.[25] Such an epokhality constitutes a shock, and for this reason we must understand technical life as a life of shocks. These moments are not dialectical because the poisonous aspect of this pharmacology is irreducible. For example, if individuation occurs as rational knowledge, this knowledge may *always* some day or other come to serve stupidity: there is no absolute knowledge.

This is because "the I or the ego," as a fantasy of identity (as the *purely psychic* as well as the *purely social*) is the point of articulation of these three dimensions of disindividuation that Deleuze calls "indices of the species" (Deleuze, *Difference and Repetition* 151).

As such, it is necessary to posit that the psychic individual is the individual who is *capable of disindividuating* (just as, according to Canguilhem, the technique of healthy living lies in "*the power and the will to fall sick*" [Canguilhem 200; translation modified]) through a disindividuation due not to his will but to the artefactual (factical) and pharmacological situation through which alone it is possible to say and do stupid things, and where *saying* here often means *doing* – stupidity being also and perhaps *especially* performative (and perhaps Gabriel Tarde should be read from this perspective).

notes

1 This was the year of publication, though it was written in 1944.

2 The spread of this word, *malin*, particularly in marketing and advertising, which initially referred to the Devil and which has come to designate cunning intelligence and a "wise guy" [*petit malin*], is a symptom typical of our misery.

3 See also Habermas: "In the transition from the literary journalism of private individuals to the public services of the mass media the public sphere was transformed by the influx of private interests, which received special prominence in the mass media" (53).

4 See Adorno and Horkheimer 214. The link made in this fragment by Adorno and Horkheimer between stupidity and frustrated desire, and which they inscribe here into a perspective that I would call organological, must be analysed as a process of the *regression of desire towards the drives*. The fragmentary and incomplete character of these notes and sketches, however, prevents going further here.

5 This is what became clear to Freud in 1920.

6 I have previously commented on this remark by Zabunyan in Stiegler, *Uncontrollable Societies* (24).

7 "Quite generally, the well-known, just because it is well-known, is not cognitively understood. The commonest way in which we deceive either ourselves or others about understanding is by assuming something as familiar, and accepting it on that account" (Hegel, *Phenomenology of Spirit* §31; translation modified). My citation of Hegel to support a proposition by Deleuze no doubt seems surprising. This is, however, precisely a question of the well-known belief, which is also to say, the stupid belief, that Deleuze – and Nietzsche – *oppose* Hegel.

8 I refer here to the height of what Gilbert Simondon called "key-points," which are culminating points, highlights, and to *those heights which mean there are base thoughts as the truths from which they are made*:

> There are imbecile thoughts, imbecile discourses, that are made up entirely of truths; but these truths are base, they are those of a base, heavy and leaden soul. The state of mind dominated by reactive forces, *by right*, expresses *stupidity and, more profoundly, that which it is a symptom of: a base way of thinking*. (Deleuze, *Nietzsche and Philosophy* 105)

In relation to the worst, and to the worst stupidity, see Sophocles, *Antigone*, and my commentary in *Uncontrollable Societies* (24ff.).

9 See Derrida, "The Supplement of Copula: Philosophy before Linguistics" in *Margins of Philosophy* 175–205.

10 I have attempted to begin this analysis in *Technics and Time, 1* and *Technics and Time, 2*. I will return to this question in *Technics and Time, 4* (forthcoming).

11 If this is so, this is because Derrida *plays* the fool (*fait la bête*, which is not necessarily the same thing as doing stupid things, *faire des bêtises*), and not because he *is* stupid [*est bête*]. Anyway, *who could* one say *is* stupid? Heidegger, for example? Surely not. Heidegger, who was not exactly stupid, who was "not just stupid" [*juste pas bête*] as the younger generations say today, *did* – that is, *said* – stupid things. And in this case he was not content to "play the fool." However that may be, in relation to stupidity, *being and not being* perhaps do not agree, perhaps never agree, even when these copulas are determined or undetermined by the adverb "exactly." Between *being* (stupid), *doing* (stupid things), *and saying* (stupid things), the question of stupidity would be at the same time older, deeper and lower than the *question of being and of spirit*, including in *Of Spirit: Heidegger and the Question*, where Derrida *approaches* the question of the animal "poor in world." The *default* of spirit, that is, the *feeling of not having any*: such would be the commencement of spirit starting from that which is stupid, *epimetheia* (and this is also *la Bête de la Belle*).

12 I will return to this question of the indeterminate in Deleuze, which must be compared to the question of the indeterminate in Heidegger – in passing through the relation to death.

13 To this must be added the process of technical individuation, which psycho-social individuation presupposes, even though Simondon is not very clear about this. See Stiegler, *De la misère symbolique 1*.

14 It is this that enables Simondon to think industry. I have tried to analyse this reversal of relations between being and possibility in *Technics and Time, 3*, in *Économie de l'hypermatériel et psychopouvoir* and in *What Makes Life Worth Living*.

15 The crystal is the individuation of an amorphous milieu from which emerges individuality, that is, a physical individual. See *L'Individu et sa genèse physico-biologique* 83.

16 I shall return to these questions and to the question of animality in *Veux-tu devenir mon ami?* (forthcoming).

17 I have tried to show that it is this logic that is at work in what Heidegger calls "*das Man*" (the *they* or the *one*). See Stiegler, "The Theatre of Individuation," and idem, "To Love, To Love Me, To Love Us: From September 11 to April 21" in *Acting Out*.

18 See Blanchot and my commentary in *Veux-tu devenir mon ami?* (forthcoming).

19 In the sense given to this in Simondon, *Imagination et invention* 13.

20 Derrida, *Margins of Philosophy* 3:

> I will speak, therefore, of the letter *a*, this initial letter which it apparently has been necessary to insinuate, here and there, into the writing of the word *difference*; and to do so in the course of a writing on writing, and also of a writing within writing whose different trajectories thereby find themselves, at certain very determined points, intersecting with a kind of gross spelling mistake.

21 And this is a trait common to both Derrida and Deleuze.

22 In his interpretation of the theory of the three souls outlined by Aristotle in *On the Soul* – where vital individuation in the Simondonian sense includes both the vegetative and sensitive stages of the soul – Hegel shows that any noetic soul (any psychic individual) can regress to an animal state. It means that they are in a deferred and suspended relation to their own possibility, held within their "in itself" without passing into the act of the "for itself." And this is not without relation to Deleuze's statement about stupidity as a form which does not take. See Hegel, *Lectures on the History of Philosophy*.

23 It is true that Simondon's thesis, from which these lines are extracted, was defended seven years before Leroi-Gourhan published *Gesture and Speech*.

24 That is, the interruption, suspension and transformation, or the individuation, of an earlier individuation.

25 On the *doubly epokhal redoubling*, see *What Makes Life Worth Living, Technics and Time, 1* and *Technics and Time, 2*.

bibliography

Adorno, Theodor, and Max Horkheimer. *Dialectic of the Enlightenment: Philosophical Fragments*. Trans. Edmund Jephcott. Stanford: Stanford UP, 2002. Print.

Bataille, Georges. *Lascaux, or The Birth of Art*. Trans. Austryn Wainhouse. Lausanne: Skira, 1955. Print.

Blanchot, Maurice. *Friendship*. Trans. Elizabeth Rottenberg. Stanford: Stanford UP, 1997. Print.

Camus, Albert. *The Myth of Sisyphus*. Trans. Justin O'Brien. London: Penguin, 1975. Print.

Canguilhem, Georges. *The Normal and the Pathological*. Trans. Carolyn R. Fawcett. New York: Zone, 1991. Print.

Deleuze, Gilles. *Difference and Repetition*. Trans. Paul Patton. New York: Columbia UP, 1994. Print.

Deleuze, Gilles. *Nietzsche and Philosophy*. Trans. Hugh Tomlinson. New York: Columbia UP, 1983. Print.

Derrida, Jacques. *The Beast and the Sovereign, Volume 1*. Trans. Geoffrey Bennington. Chicago and London: U of Chicago P, 2009. Print.

Derrida, Jacques. *Margins of Philosophy*. Trans. Alan Bass. Chicago and London: U of Chicago P, 1982. Print.

Freud, Sigmund. *Civilization and Its Discontents*. Trans. James Strachey. *The Standard Edition of the Complete Psychological Works of Sigmund Freud*. Vol. 21. London: Hogarth, 1964. Print.

Habermas, Jürgen. "The Public Sphere: An Encyclopedia Article (1964)." *New German Critique* 3 (1974): 49–55. Print.

Hegel, Georg Wilhelm Friedrich. *Lectures on the History of Philosophy, 1825–6*. Vol. II. *Greek Philosophy*. Trans. Robert F. Brown. Oxford: Oxford UP, 2006. Print.

Hegel, Georg Wilhelm Friedrich. *Phenomenology of Spirit*. Trans. A.V. Miller. Oxford: Oxford UP, 1977. Print.

Heidegger, Martin. "The Self-Assertion of the German University." *Martin Heidegger and National Socialism: Questions and Answers*. Ed. Günther Neske and Emil Kettering. Trans. Lisa Harries. New York: Paragon, 1990. 5–13. Print.

Kant, Immanuel. *Groundwork of the Metaphysic of Morals*. Trans. H.J. Paton. New York: Harper, 1964. Print.

Lacan, Jacques. *Écrits*. Trans. Bruce Fink. New York and London: Norton, 2006. Print.

Marx, Karl, and Friedrich Engels. *The Communist Manifesto*. Trans. Samuel Moore. London: Penguin, 1967. Print.

Nietzsche, Friedrich. *The Gay Science*. Trans. Walter Kaufmann. New York: Vintage, 1974. Print.

Nietzsche, Friedrich. *Human, All Too Human*. Trans. R.J. Hollingdale. Cambridge and New York: Cambridge UP, 1986. Print.

Nietzsche, Friedrich. *On the Genealogy of Morality*. Trans. Carol Diethe. Cambridge and New York: Cambridge UP, 1994. Print.

Nietzsche, Friedrich. *Thus Spoke Zarathustra*. Trans. R.J. Hollingdale. London: Penguin, 1961. Print.

Simondon, Gilbert. *Du mode d'existence des objets techniques*. Paris: Aubier, 2001. Print.

Simondon, Gilbert. *Imagination et invention*. Chatou: Transparence, 2008. Print.

Simondon, Gilbert. *L'Individu et sa genèse physico-biologique*. Grenoble: Millon, 1995. Print.

Simondon, Gilbert. *L'Individuation à la lumière des notions de forme et d'information*. Grenoble: Millon, 2005. Print.

Simondon, Gilbert. *L'Individuation psychique et collective*. Paris: Aubier, 2007. Print.

Stiegler, Bernard. *Acting Out*. Trans. David Barison, Daniel Ross, and Patrick Crogan. Stanford: Stanford UP, 2009. Print.

Stiegler, Bernard. *De la misère symbolique 1. L'Époque hyperindustrielle*. Paris: Galilée, 2004. Print.

Stiegler, Bernard. *Économie de l'hypermatériel et psychopouvoir*. Paris: Mille et une nuits, 2008. Print.

Stiegler, Bernard. *Taking Care of Youth and the Generations*. Trans. Stephen Barker. Stanford: Stanford UP, 2010. Print.

Stiegler, Bernard. *Technics and Time, 1: The Fault of Epimetheus*. Trans. Richard Beardsworth and George Collins. Stanford: Stanford UP, 1998. Print.

Stiegler, Bernard. *Technics and Time, 2: Disorientation*. Trans. Stephen Barker. Stanford: Stanford UP, 2009. Print.

Stiegler, Bernard. *Technics and Time, 3: Cinematic Time and the Question of Malaise.* Trans. Stephen Barker. Stanford: Stanford UP, 2011. Print.

Stiegler, Bernard. "The Theatre of Individuation: Phase-Shift and Resolution in Simondon and Heidegger." *Parrhesia* 7 (2009): 46–57. Print.

Stiegler, Bernard. *Uncontrollable Societies of Disaffected Individuals: Disbelief and Discredit.* Vol. 2. Trans. Daniel Ross. Cambridge: Polity, 2013. Print.

Stiegler, Bernard. *What Makes Life Worth Living: On Pharmacology.* Trans. Daniel Ross. Cambridge: Polity, 2013. Print.

Zabunyan, Dork. "L'Apprentissage de la bêtise." Paper delivered at the "Deleuze" conference, ENS Ulm, 2004. Unpublished.

steve baker

FIVE HERALDIC ANIMALS (FOR EDUARDO KAC)

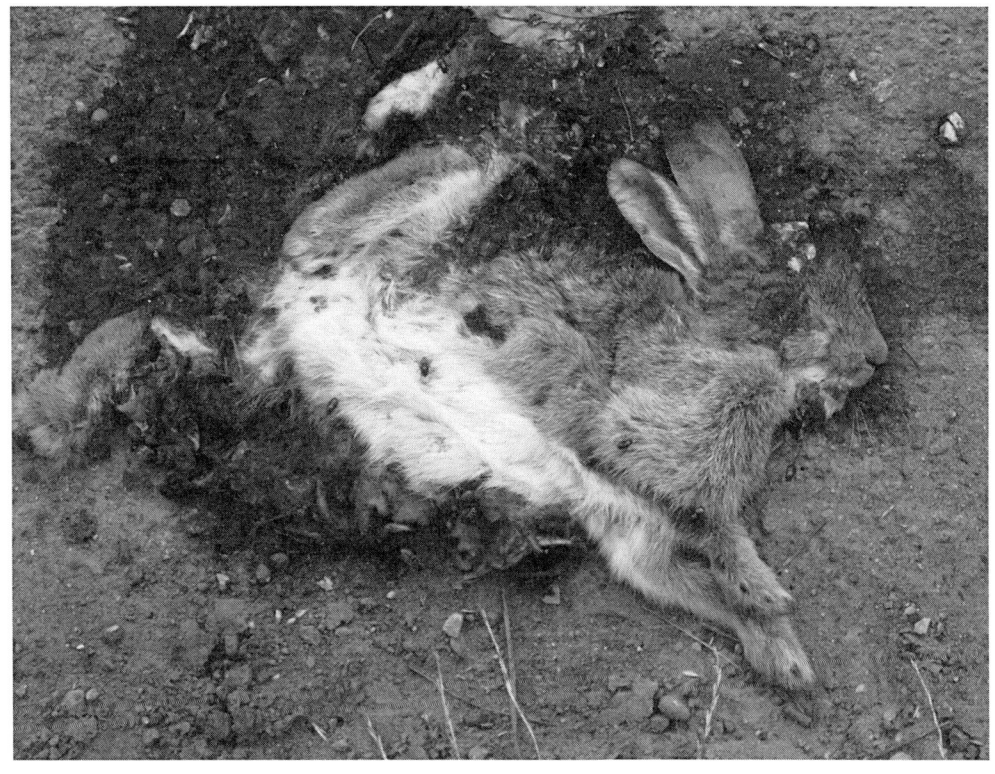

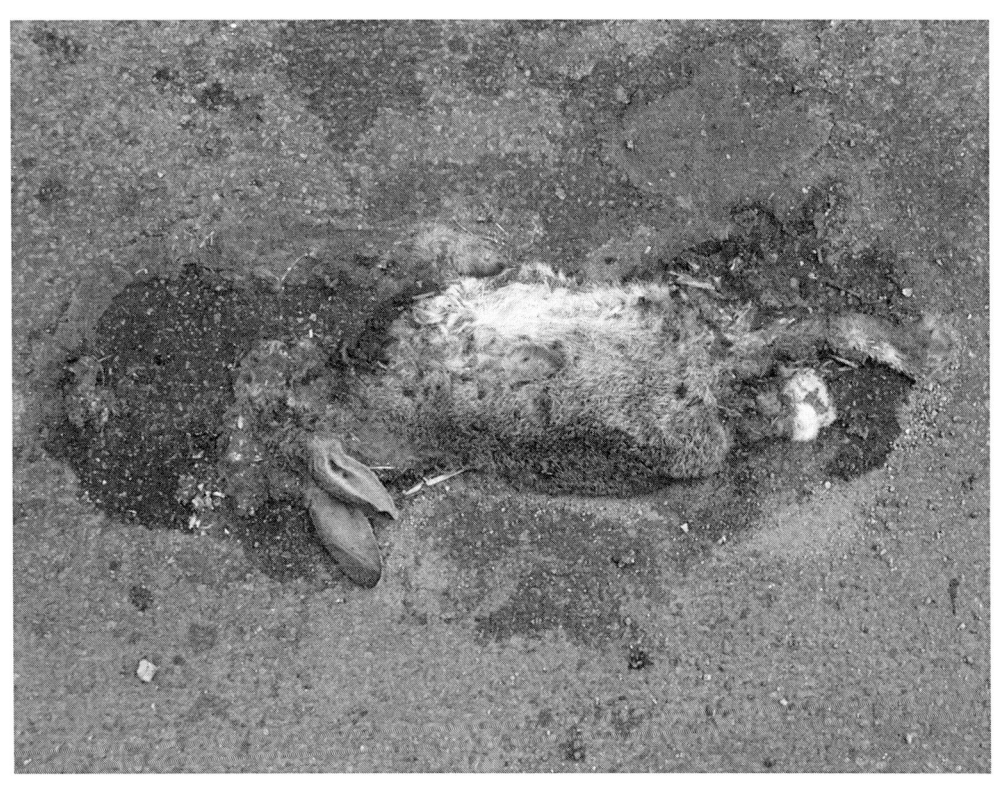

Ron Broglio: In Philosophy and Animal Life *you along with Cora Diamond open up a shared difference with other animals. We along with other animals are exposed. We house within us a vulnerability co-substantial with corporality and liveness. Biopolitics and dispositif leverage corporality toward a way of managing lives. But Foucault also sketches a line of thought by which biopower is that which cannot be assimilated. For example* homo oeconomicus *in which "Economic rationality is not only surrounded by, but founded on the unknowability of, the totality of the process." Moving from a restricted to a general economy, can animal and human bodies, vulnerable bodies, produce an unknowability over and against the frames and apparatus of biopolitics? Which is to say, how can that which is vulnerable also take us elsewhere?*

ron broglio

AFTER ANIMALITY, BEFORE THE LAW
interview with cary wolfe[1]

Cary Wolfe: Yes, that's something I'm very attracted to in Foucault's work. This is something that Maurizio Lazzarato's reading of Foucault emphasizes, and something that Jeff Nealon emphasizes in his book on Foucault as well: that biopolitics is always a somewhat fantasized reduction of a more complex relation to biopower, fantasized because it wants to always control something that's inherently risky. So, when Foucault talks about power and resistance being in this permanently unstable relationship, the way I think about this is that with the shift to biopolitics the body in the most general sense becomes a new political resource. But to have the goodies that go with this new political resource, certain risks are entailed, right? Certain relations of alterity are entailed and levels of unexpectedness are entailed, and

that's exactly what generates the new forms of surveillance, and control, and micromanagement, and the new forms of power/knowledge that attempt to deploy this new political resource and not get burned by the resistance and alterity that's always a part of the transaction.

RB: So, the friction of the bodies? A vulnerability, flesh, and exposure.

CW: That's right. And so one way to think about it is that the relationship between the biopolitical proper and biopower is always a dynamic and risky – and, above all, a strategic – relation in ways that biopolitics cannot always completely foresee. And so, what

Esposito's language would call a more "affirmative" way of understanding biopolitics emphasizes this space of resistance that is part of the gambit of biopolitics itself. So even in the most oppressive and dominating forms of the biopolitical use of biopower, there is always a potentially aleatory and creative element that can erupt in very unexpected ways at very unexpected times. To me, this is a much more robust understanding of the biopolitical than you would get from Agamben or the Schmittian line, with its emphasis on sovereignty and the abjection of the body as "animal," which in turn becomes a kind of abstract philosophical *topos*. What we're talking about here instead is a kind of biopolitics that is very, very specifically articulated in relation to different, particular kinds of bodies.

RB: Right, so their resistance or how they play out is a matter of risk?

CW: That's right – it's risky. Biopolitics can't have the goodies of the new political resource without running the attendant risk. And that's how Foucault's rendering of biopolitics enables us to rearticulate questions of political effectivity in the relationship of biopower to biopolitics.

RB: Is this discussion of vulnerability, flesh, and exposure inscribed within phenomenology and so still within the purview of the human? One could trace these concepts to Merleau-Ponty's flesh of the world and to his work on chiasm. Or is it possible that vulnerability, flesh, and exposure are a leap from phenomenology to something else – a posthuman phenomenology? Here I am thinking of Merleau-Ponty's unfinished work The Visible and the Invisible *in which opacity or invisibility is not a consequence of our failed sense organs but rather the nature of things in a world of perception. Withdrawal or invisibility is an essential part of the being of beings. Here, then, Heidegger's figure of the human as shepherd of being lags the very beings he intends to shepherd.*

CW: Yes, I think that's really well put. And this is what Esposito is attempting to do with this figure of flesh in his own work on the

biopolitical. I talk about this some in the book: that biopolitics actually operates more precisely at the level of flesh than the level of "the body." Flesh is a shared kind of organicity and a shared kind of embodiedness for which "the body" then becomes a kind of foreshortening and, already, a kind of closure. Having said that – and this is something I insist on in the book – I don't want to evacuate questions of the phenomenological with regard to specific forms of life and specific ways of being in the world. This is one reason I return to and try to retheorize the concept of Dasein and try to radicalize, actually, what Heidegger was right about: that Dasein has this radically inhuman, ahuman, character, or as he puts it, that the being of beings is not a being. I try to rearticulate the radically ahuman character of Dasein as consisting of a constitutively prosthetic relationship between the biological wetware of specific types of life forms and the exteriority of technicity that rewires that wetware. That technicity doesn't need to be tools; it can simply be the semiotic in the most bare, fundamental sense. And so some creatures (but not all) have this constitutively prosthetic relationship to technicity and exteriority in a way that others don't. For example, that can't happen for a cockroach in the same way because a cockroach has an exoskeleton that limits its ability to have a sufficient concentration of neural tissue, which in turn limits the kind of plasticity that can be reconfigured in relationship to this exteriority of technicity in a way that's different from a dolphin or an elephant. That doesn't mean that we can really be certain about drawing "a line," and on one side of that line we would find Dasein and on the other side of that line we wouldn't. In fact, the history of science clearly shows that that line is constantly moving and constantly shifting – under erasure, you might say. I mean, look at what we've learned about certain kinds of animals over the past fifty years. So it's not an issue of drawing a line; it's an issue of articulating the specificity of how different forms of life exist in the world, and the ethical and political consequences for norms that one would want to derive from that. And insisting on that phenomenological dimension is precisely what Heidegger is right about critiquing biological or zoological continuism; he's right that these kinds of issues can't be simply reduced to a kind of zoological or biological description. What he's wrong about is that the radically ahuman character of Dasein coincides in any way with the distinction between human and animal. And so I think it's crucial to hold onto the phenomenological dimension even as you can give a robust naturalistic, rather than transcendental, account of how the phenomenological arises in the first place.

You could point to what I call in the book the scandal of the cephalopods, for example. The scandal of the cephalopods is that cephalopods demonstrate everything that I mean by nonhuman, animal Dasein. And yet, they don't have the physiological and biological architecture that is shared with mammals – they're not vertebrates or even chordates, number one. And number two, the kinds of behaviors that they demonstrate have always been presumed to exist only in animals that have a sufficiently long lifespan and live in social groups. But neither is true for the cephalopods. Giant squids live about three years ...

RB: Wow, I didn't realize so short.

CW: Yes, and almost all of the cephalopods are solitary.

RB: Yes.

CW: So, where does this Dasein come from? Nobody knows.

RB: We were talking about Bateson earlier, and his "ecology of mind" stuff seems useful here.

CW: Yes. To fully evacuate the humanism of the description, you have to look back into a kind of systems theoretical description of autopoiesis, emergent behaviors, structural couplings, embodied enaction, and the like to have a vocabulary that enables you to talk about the specificity of these things in a robust way across species lines. And it enables you to give – and this is, to me, the theoretical challenge that has to be met and is difficult – it enables

you to give a robust, naturalistic account of how phenomenological domains arise that cannot be reduced to the biological or the zoological. At this point, then, you could also veer off, if you wanted, into Stiegler's entire rereading of Heidegger, which would be an important part of that conversation. But that's the articulation it seems to me that it has to be made in relation to reframing the specific character of the phenomenological.

RB: That's interesting because then once you play that out you begin to talk about cultures, right?

CW: Very quickly you do. And even there I think it's useful to more finely parse the vocabulary. I'm thinking, for example, of Maturana and Varela's discussion of the difference between linguistic domains and language proper. Those are actually two different things, and language proper is actually a second order phenomenon that arises on the basis of the capacity for linguistic domains that is broadly shared by lots of different nonhuman animals. But that actually only leads to language proper in only very rare instances. It's a much more finely tuned theoretical apparatus, I think, for describing what we would otherwise call phenomenology.

RB: That's right. So phenomenology itself is too much of a blanket term?

CW: Far too much of a blanket term. It's too blunt of an instrument. Even as the insistence on disarticulating the properly phenomenological from the biological and the zoological is absolutely right. That's why Derrida himself agrees with Heidegger's resistance to "biological continuism," and says it would be "asinine."

RB: So, this is slightly along the lines of Derrida in The Animal That Therefore I Am *where he develops contours by which to differentiate a community of beings "not in affecting the limit, but in multiplying its figures and complicating, thickening ..." The multiply folded, re-fractured abyss relates to what you are doing in the middle of your book where you address the problem of welcoming all animals and the problem of infinite hospitality. I think you're responding to Esposito's call for recognition of "life singular and impersonal." Hospitality, to such a wide swathe of life, and forms of life, and living beings, seems impossible or an ideal limit – an absolute hospitality.*

CW: Right. That's the passage I was invoking.

RB: So it's not a matter of division between man and animal, but rather humans among other animals differentiated by a number of figures and folds. So in all of this, then, how is one to think about the limits of hospitality between humans and other animals?

CW: I think the sense of this that's underneath your question is spot on, and this is related to something I was talking about earlier. What Esposito is up to at the end of *Bios* seems to me, actually for this very reason, a radically *de*-differentiating move in declaring in effect the radical equality in all forms of life. Now, that's problematic for a number of reasons, some of them philosophical and some of them pragmatic – I mean since we are, after all, to take him at his word that we are talking about norms in relation to ethics and politics. A pragmatic problem that I underscore in the book is that his position takes us right back to the debates that took place in North America, but beyond as well, back in the 70s and 80s around biocentrism and around deep ecology. So, if he's right, are we then supposed to allow anthrax, and Ebola virus, and Hanta virus, and S.A.R.S., and so on, to achieve their creative flourishing even if it means, you know, a 70 percent die off of the human population? You're very quickly led to these kinds of pragmatic dilemmas. And that's a debate that I don't think Esposito or people working in Italian political philosophy around biopolitics really have any reason to know much about – the debates about deep ecology and biocentrism. A theoretical and philosophical problem – and this is the part that's related to what I was talking about earlier – is that it's not as if the organism is ever in a position to unilaterally exercise an unconditional hospitality, either

biologically, or physiologically (or even, you might say more broadly, socially or culturally). In other words, whatever that process of extending hospitality is, it radically has (to use psychoanalytic language) an unconscious, one that's not just psychoanalytic but is actually biological, physiological, and involves communities of different kinds of life forms that inhabit different creatures (and vice versa) in radically different ways in terms of their own autopoiesis. And so the point of unconditional hospitality is not that it's actually possible, but that it should always be a kind of horizon to strive for to prevent the immunitary foreclosure of self-reference. This horizon, or this desire, or whatever you want to call it, of unconditional hospitality, is precisely to remind you that in that decision and in that moment you will be shown, at some point later, to have been wrong. And the reason that this is crucial to biopolitical thought is that it keeps that zone of immunological protection from automatically turning into, as Derrida puts it, an autoimmune disorder. The idea is that once you start drawing lines between humans and animals, Aryans and Jews, Muslims and Christians, that is always going to lead to the runaway train process of an autoimmune disorder. So eventually, you know, how Aryan is Aryan enough? How Christian is Christian enough? How human is human enough? How "proper," to go back to Heidegger, is proper enough? The horizon of unconditional hospitality as something to strive for is precisely calculated to remind you that whatever those liens are that you are drawing have to always be taken under erasure, *even as*, pragmatically, those lines have to be drawn and *are* drawn all the time. As Derrida puts it, in offering any kind of hospitality you are always performatively offering something specific, determinate, and conditional. If you weren't, things like politics would have no binding force.

RB: For Heidegger, animals are benumb to the possibility of seeing objects themselves? The animals do see and interact with the world around them. So, if not a Heideggerian clearing that's reserved, say, for humans, then is it possible that animals have some other sort of clearing and revealing available to them and not to humans? So even Heidegger, in addressing the poverty of animals, admits this. That is, the living flesh of animals is not "something inferior, or that is a lower level when in comparison to the human Dasein. On the contrary, life is a demanding which possesses a wealth of being opened, of which the human world may know nothing at all." So is it possible to think of such a wealth of being open? And how would one approach such a mode of thinking?

CW: It is attractive to think that, but I think it requires an articulation that's radically non-Heideggerian. And the reason I think it has to be radically non-Heideggerian is that the distinction between the open and its other is like the distinction between the "proper" and the improper in Heidegger. It is a firm distinction, and a distinction that is not under erasure, as Derrida would put it. And actually, what I argue – quite the contrary – in the book is that Heidegger's description of animals having a world in the mode of not having is maybe the best description of Dasein that we're going to get. *Humans* don't have a relationship to the world, in Heidegger's words, "as such and in its being." That's a fantasy of the "proper" of the human in relation to the open that's made possible by the fact that humans, for Heidegger, possess language and animals don't, which in turn grounds the phenomenological possibility to have a relation to the world "as such and in its being." What I argue (and this is really just following Derrida's reading of Heidegger) is that human beings don't have a relationship to the world as such and in its being because language doesn't work that way and meaning doesn't work that way. So our relationship to the world through language or through other forms of meaning, or other sorts of knowledge-making schemata, is made possible by the technicity of a semiosis that is radically *not ours* and radically *not us* – and, in fact, radically nonorganic. It is machinic, or *une "grille,"* as Derrida puts it in *Signature, Event, Context.* And so what is most ours, what is most "proper," to our relationship to the open is precisely that which is least ours

and least proper. It's a polluted, contaminated, hybrid relationship between the human and the nonhuman, the biological and the technical, that makes possible any apprehension of the world whatsoever. So the funny thing is that at the end of the day, to make a very long story short, Heidegger's description of animals in relation to the world as "having a world in the mode of not having" is actually the best description of Dasein that we have.

RB: You know what's interesting is very early on you were saying we really need a different language because almost formed within the Heideggerian construct he's already prefigured his own sense of a humanity as a very sort of particular way of being.

CW: Yes, and Derrida's emphasis on this question of the "proper" in Heidegger is a key moment in drawing that out.

RB: So what sorts of language do you think will become available for talking about this?

CW: Well, I think there are different languages that can be useful here. One that I've already mentioned in my own work is my attempt to cross articulate the kind of work that Derrida does with work in second order systems theory and, beyond that, in people like Bateson. But another example – just to sort of shift gears for a moment – that I know you're really interested in is the kind of work that artists do. Artists have a lot to teach us about the relationship between specific sensory modes that are different for different creatures in relation to the media that make possible specific forms of meaning. They're sensitive to it because they work with those media so closely, in a way that people who just work with words and texts aren't. So I think there are different kinds of languages and different kinds of practices that can address these questions. Something that needs to be added here (and this is maybe just a segue into talking about the Posthumanities series) is that this doesn't mean that the traditional languages of humanism aren't useful and important. This goes back to what I was saying earlier about not wanting to jettison what Heidegger's right about, I think, in the question of the phenomenological. It would be a humanist fantasy, in fact, to think that one could just "post-humanistically" institute a radical rupture and not be informed by those languages. It's more about saying, well, you know, the commitments and the desires and the interests of a lot of work in humanism are important. But the philosophical and theoretical tools used to articulate those commitments are actually self-defeating and actually undercut and foreclose those interests and commitments. An example of this that I discuss in *What is Posthumanism?* is the use of rights discourse in relation to the ethical standing of nonhuman animals. I think we would all agree that an admirable desire of humanism would be to respect the standing of at least some nonhuman animals and to protect them from exploitation, cruelty, and so on. But the attempt to articulate that desire, which is an admirable one, in terms of the rights framework ends up foreclosing and undercutting that desire by reinstating a normative picture of the subject of rights that ends up being humanist and anthropocentric through and through, that ends up with a being that looks a lot like us, so that, in the end, nonhuman animals matter because they are just a diminished version of us. It seems to me self-evident that trying to think about the value of dolphins in terms of their being diminished versions of *Homo sapiens* makes no sense. And at that point, people like Judith Butler and Cora Diamond, on these issues of vulnerability and precarious life, come back into the picture as an alternative to the entire rights framework.

RB: You know, I think the ability to think the vulnerable ... I mean reason is such a tool of mastery and so many of our projects have been projects of mastery; so, to think the vulnerable and to write it in a way that itself is not necessarily masterful. That is, you're not mastering vulnerability either. It is sort of a current challenge.

CW: Did you see this film *Buck*?

RB: No.

CW: This is something you should see. It's a documentary about this real Montana cowboy who is sort of the original "horse whisperer."

RB: I've heard about this, yes.

CW: Monty Roberts was the first famous "horse whisperer" on the scene, but this guy, Buck, has a remarkable technique when working with horses. If you've read Vicki Hearne's work, for example, you get a picture of the super specific way that horses are in the world – their bodily sensitivity, how they're not like dogs, and so on. Anyway, the interesting thing about this documentary is that this guy does a kind of traveling road show, and he goes into these highly skeptical communities where the only way people know how to train horses is by "breaking" horses. Well, he goes into these communities and in about ten or fifteen minutes the traditional horse breaking people on the scene watch what he's doing with the horses and they just can't believe it. As he's interacting with the horses, he's taking advantage of his understanding of their visual field and how they see the world, how they respond to different kinds of touch, both light and heavy, all sorts of things. Like Monty Roberts, he has an extraordinary sensitivity to what you might call the bodily gestural nature and repertoire of horses and how it is constantly communicating in ways that he's sensitive to but other people can't see. Anyway, as the story unfolds, we learn that when he was a child he and his brother would go on television and perform all these roping and lassoing tricks. They were little cowboy celebrities. But what Buck also reveals, as the movie unfolds, is that he and his brother were severely beaten by their father when they were kids – beaten and almost "broken," really. So what becomes clear in the film is that there is a very deep connection between his own experience of his own bodily vulnerability as child, his own bodily exposure to violence, to what his dad did to him, and his refusal to impose that sort of violence on animals, which then led him to develop great sensitivity to the embodiment of other creatures in ways that enabled him to work with horses in a way that almost no one else can. And I would say in that situation that you have in play both the embodied, the biological or zoological, and the phenomenological, in the sense of shared meaning taking place across species lines. They're both in play in that relationship. And this brings us back to these "aleatory" and creative elements of biopower that we were talking about earlier.

RB: Let's turn the conversation a bit. I'd like to talk some about your book series Posthumanism with the University of Minnesota Press. By now the series has had enough books in it that you're beginning to get a sense of the mosaic regarding what is this question of posthumanism and how it's developing. Maybe we can talk a little bit about that as well as the types of books in the series.

CW: The original rubric in the series as I presented it to the press has to do with posthumanism as a fundamental decentering of the human as a constitutive category, one that can't, any more, do any heavy lifting in an explanatory way in relation to the situation in which we find ourselves. You can describe that as a historical phenomenon if you want, having to do with a whole range of developments in technology, information, medicine, economics, and so on – all sorts of factors that have made the human no longer "master in his own house." The series rubric was originally framed as exploring that question between two poles. One was the decentering of the human vis-à-vis a "green" dimension, in relation to questions of ecology, so-called animal rights, environmentalism, global warming, the extinction of other creatures, etc. The imbrication of the human within these larger biological, ecological, zoological networks of relations. The other pole, the so-called "gray" pole, would be the decentering of the human in relation to informatic and communication technologies, economic structures and systems, and different forms of media in an increasingly prosthetic and over-determined relation of the human and its ways of knowing and acting in the world to things like cell phones or the Internet. When projects come to the series, they tend to break off in one of those two directions. For example, Mick Smith's book on ecological sovereignty is an obvious example; your book would be another example that's clearly associated with the

"green" pole – our relations to nonhuman animals and more broadly the environment. But then we have other works in the series that gravitate more toward the "gray" end of the spectrum. Thierry Bardini's book *Junkware* would be a good example. But a lot of books in the series actually work at the crossing of these two as well. Dominic Pettman's book *Human Error: Species Being and Media Machines* is a good example. And, you know, thinking back to what you might call the prehistory of the series, Donna Haraway's work has always been in a way situated at that crossing point. So those are the poles that define the nexus within which the series operates, and when you think about it, it is a pretty generous space. What I wanted to do in the series is that I precisely *did not* want to do an "animal studies" series. As I argue in *What is Posthumanism?*, I don't even like this term "the animal," much less the term "animal studies," and even if I did, both of those for me are subsets and sub-questions of this broader question of posthumanism. "The animal" is one terrain on which those questions of posthumanism play themselves out and can be explored with particular stakes that are different, of course, from other areas that you might investigate. So my hope is that these two poles open up a space in which these constitutive terms can constantly destabilize each other and trouble each other and interrogate each other to deepen our sense of the ways in which posthumanism plays out in different areas. A major challenge of the series that we were talking about earlier is, you know, that a book series is something that unfolds in real time, so you don't want to do four or five of the same kind of book at the same time. It's not just about the abstract, synchronic, conceptual nature of the series; you're also always dealing with questions of a kind of choreography of the list for the fall and the spring, the mix of books that are going to be on that list. We get projects that are right for the series, but I'll have to tell authors that we just did a book like that, or we're about to. So that's a second level on which the mosaic is constantly reconfigured. My feeling is really we're just getting started.

The series began, technically, in 2007, and we were doing four books a year – now we're doing six books a year. My challenge as an editor is to keep the series unexpected and fresh for people, and not let it turn into something – and this a danger for any book series or a journal – that becomes a mechanism for codification of a particular position or theoretical approach. So, for example, we're doing a clutch of books right now around Speculative Realism and Object Oriented Ontology – Ian Bogost's book *Alien Phenomenology*, among others. This is not because I myself am especially enamored of Object Oriented Ontology or Speculative Realism – I'm not. It's that I think there are a lot of obviously very smart people who are interested in this work right now, so perhaps if that work is a part of the series, there can be some interesting crosstalk between it and other, very different, work in the series; it'll make possible a conversation that maybe I myself am not that interested in getting that deeply into in my own work, but that many readers will be interested in. So that's the editorial challenge: how to use your own expertise, but not let it turn into a matter of simply reproducing your own thinking, your own approach to these questions.

RB: Yes. Do you think that having seen the series play out and also work that's in the pipeline that your understanding of posthumanism has changed as a result of that?

CW: Yes, I think it has changed because just in the five years that the series has been up and running, a lot of the conversations have changed. Take, for example, my own work and the new book that we've just been talking about. Five, six, seven years ago, questions of biopolitics were really not that central to my approach to these questions around nonhuman life. That's different now, and it's different partly because of work that we've done in the series, projects that I've read in consideration of the series, and conversations that editing the series made me aware of in a deeper way than I would have been otherwise. So, for me, part of the payoff of editing the series is that I'm constantly being reeducated by the projects that come across my desk, which in turn reshape how I think about these questions. So the biopolitical dimension of the issue is central for me now in a way that it wasn't five or six years ago, in no small part because of the work I've been doing as an editor.

RB: So, do you think in Foucault's sort of understanding of the biopolitical, that there is already something of the posthuman within it?

CW: Absolutely, yes. This goes back to what we were talking about earlier. The posthumanist dimension – or one of the posthumanist dimensions, I actually think there are many – of Foucault's work on this question is precisely to demonstrate that whatever we're talking about here is not a question of "the subject." That is not an adequate vocabulary to describe the highly complex articulations of bodies – human and nonhuman – with technologies in relation to political and economic formations. Foucault allows us to rearticulate these questions so that, as I argue in the book, you can understand something like factory farming – and the fallout of factory farming, economically, ecologically, also with regard to things like antibiotics and public health – as what Foucault calls "a new schema of politicization." It's not just an ethical issue. It's a robustly political issue. But you've got to have a political vocabulary that allows you to talk about the articulation of technologies and bodies in the service of the commodification of life in a way that makes it legible as a political issue. So that requires a posthumanist vocabulary that sees politics as no longer being about questions of agency, autonomy, intentionality, the subject, reciprocity, and so on, but something far more complicated and distributed going on.

note

1 The following interview was conducted while at the Modern Language Association conference in Seattle. The conversation took place on 14

January 2012 at the Edgewater Hotel. Wolfe's book *Before the Law: Humans and Animals in a Biopolitical Frame* was still forthcoming and discussion of the book was from a completed draft.

Index

INDEX